D0674147

CHILDREN OF LIGHT

Also by Gavin Weightman

The Industrial Revolutionaries
London River: A History of the Thames
The Frozen Water Trade
Signor Marconi's Magic Box

CHILDREN OF LIGHT

HOW ELECTRICITY CHANGED BRITAIN FOREVER

Gavin Weightman

Atlantic Books
London

First published in Great Britain in 2011 by Atlantic Books,
an imprint of Atlantic Books Ltd.

1 2 3 4 5 6 7 8 9

A CIP catalogue record for this book
is available from the British Library.

ISBN: 978 184887 225 7

Printed in Great Britain

Atlantic Books
An imprint of Atlantic Books Ltd
Ormond House
26–27 Boswell Street
London
WC1N 3JZ

www.atlantic-books.co.uk

CONTENTS

ACKNOWLEDGEMENTS

In 1941, the historian Marc Bloch wrote: 'In reproaching "traditional history", Paul Valéry has cited "the conquest of the earth" by electricity, as an example of one of those "notable phenomena" which it neglects, despite the fact that they have "more meaning and greater possibilities of shaping our immediate future than all the political events combined".' The French philosopher Valéry thought this inevitable, as there was not much material for the historian to explore. Bloch disagreed: surely there were copious company reports and specialist publications to be mined? The fault was not a lack of raw material but a lack of interest in technology among mainstream historians.

My experience in researching the history of electricity in Britain supports both views: there is precious little written in mainstream history on the subject, while there is a wealth of material on the rise of the industry in specialist libraries. The trouble is that the bulk of the historical material is dry and technical and lacks human interest. Some of the early memoirs are fun, but the later ones are dull as ditchwater, managerial rather than adventurous. I am therefore especially grateful to a small number of specialists who have helped me put narrative meat on the bare bones of the technical literature.

First, Dr Brian Bowers, formerly of the Science Museum, and author of many of the most authoritative and accessible books on the history of electricity and lighting, has been a very helpful guide to the available historical material. Professor John Wilson of Liverpool University, the biographer of Sebastian Ferranti, was very generous with his time in answering my questions about the life of one of the giants of electrical history. Mike Hearn was very helpful in sorting out some technical issues and was a mine of information on early electricity schemes. Peter Lamb and Andrew Smith of the South Western Electricity Historical Society shared memories of the days of the nationalized industry and gave me some valuable information on early schemes.

Asha Marvin and Sarah Hale, archivists at the IEE Library in Savoy Place, provided me with much detailed information on the Electrical Association for Women, as well as on many social aspects of the impact of electricity. All the staff of the IEE Library assisted in unearthing files from the depths of the storage system, including an official account of the impact of the Blitz on electricity supplies. At the London Transport Museum, Caroline Brick provided information on the speed limits of electric tramways, and Dick Swann of the Newcomen Society provided me with an account of an early electrical installation in London.

I had a great deal of assistance from local libraries in my research into the early days of electricity supply. Thanks to the staff of Godalming Museum, who made me welcome and provided me with a number of valuable files. Also to Lynn Humphries at Sheffield Local Studies Library, who sent me newspaper articles and other material about the world's first floodlit football match, played at Bramall Lane in 1878. Tyne and Wear Archives Service provided copies of some rare documents, notably the memoirs of Henry Edmunds, who had so many adventures working with Joseph Swan. Thanks to them and to Andrew Frost of John Frost Newspapers, who produced a valuable file on the opening of the world's first commercial power station at Calder Hall. I would like to thank Stuart Oliver of UK Coal for providing me with some valuable background on the recent history of the mines, and Isobel Rowley for introducing me to the workings of the National Grid. Dr Marjory Harper at the University of Aberdeen provided some useful references on hydroelectricity in the Highlands. Darwin Stapleton in America kindly forwarded some work he had undertaken on the electrical pioneer Charles F. Brush.

As always the staff of the London Library were extremely helpful. I would like to thank them as well as the staff of the British Library and Simon Blundell, Librarian of the Reform Club, for their help. Finally I am indebted to Angus MacKinnon at Atlantic for commissioning the book, to Annie Lee for her meticulous editing and to my agent Charles Walker and his assistant Katy Jones for looking after my interests, as always.

LIST OF ILLUSTRATIONS

INTRODUCTION

"WHAT WILL HE GROW TO?"

> With a fanfare of trumpets, the Electrical Age has arrived. At least, the electrical advertisements say so. Great central generating stations have been built and equipped with expensive machinery; steel pylons have been erected all over the country to carry electrical energy to houses, factories and farms. Only one thing is lacking. There are not enough people who are willing to buy the electricity that is produced; and those who do buy it do not buy enough. This difficulty was not anticipated by the electricians. Yet if they are to justify the millions of money spent on the Electricity Grid they must induce more and more people to buy the electricity they make.[1]

Those were the days. The quote is from a pamphlet published in 1934 with the intriguing title *The Lure of the Grid*, in which the author, Eileen

Murphy, warns women not to be seduced by propaganda encouraging them to fill their homes with electric gadgets. 'It must not be thought that I am a rabid opponent of electricity,' she assured her readers. 'On the contrary, I recognise that there are ways in which electricity is a vital part of modern life – for instance, in the supply of power to workshops and factories and for decorative "Neon" lighting. On the other hand, I dislike any attempt to stampede the housewife by extravagant promises into buying electricity for household purposes in which it has not been proved to be either economical, healthy or efficient.'

Now that Britain is thoroughly electrified, every quarterly bill arrives with advice on how to *reduce* our use of it. At some point in the history of the electrification of Britain the exhortation to use more of the stuff which so infuriated Mrs Murphy was turned on its head and we were instead asked to switch off and save wherever possible. This is symptomatic of the fact that there is something contrary in the nature of electricity which makes its history very difficult to grasp, and is perhaps the reason it barely gets a mention in most accounts of the creation of modern Britain. It is simply taken for granted, and the era, not so very long ago, when lighting a room with a flick of a switch was regarded as a great luxury is more or less forgotten. Indeed, the little filament light bulb, which first made this instant illumination possible in the early 1880s, is now to be banned because it is too power-hungry. Yet it was that invention on which the early electricity industry was founded.

Children of Light puts into its historical context the lively contemporary debate about how much electricity we should be using and how it should be generated. For at least a century after the foundation of the industry in the 1880s the problem was to keep the price down so that a larger number of people could to afford to buy it. Even in the 1950s gas lighting was not uncommon in Britain: my own primary school in north London still had gas lamps and an open coal fire in the Coronation year of 1953. Many rural parts of Britain were poorly served then, and the failure to bring electric power to the farm was regarded as something of a scandal. The reason for the delay was simple: the cost of connecting a lone building a long way from the nearest source of power could never be covered by the sale of electricity to the owner. In the poorer districts of towns the cost of connection to a power supply was too much for many people and hardly worth the expense for landlords.

The building of large power stations and linking them with a grid of high-voltage cables carried on steel towers lowered the price of electricity dramatically from the 1930s onwards. By the 1960s not many places in Britain were without electric power, and the electrification of the home began in earnest as the suburban luxuries of the thirties – vacuum cleaners, washing machines and even refrigerators – appeared in nearly all households. The manufacture of these 'labour-saving' devices and modern conveniences was often American, for the United States had stolen a march on Britain from the earliest days of the electricity industry.

Though there is still some debate about why the electric revolution came relatively late to Britain, lack of scientific and technical expertise can be discounted. After all, it was Humphry Davy who was one of the first to demonstrate how to get light from an electric charge, and his protégé Michael Faraday who devised the first electric generator. The French and Americans gained a small lead in the 1870s, but some of the earliest experiments with electric lighting were made in Britain and there were engineers anxious to exploit the new technology as well as people prepared to invest in it.

The leading innovators of the day fought bitter battles about the best way to employ electric power. However, whereas scientific dispute had the effect of resolving problems and taking the technology forward, social and political forces arose which set up a fierce resistance to the spread of electricity. Who should have the right to generate it and to sell it? Was it safe? Could it ever compete with gas for lighting? In Britain these questions, hotly debated in the newspapers and in Parliament, stifled the industry. The United States and Germany, just beginning their industrial take-off in the 1870s, embraced the new technology more eagerly. They led the way in the mass production of electric motors and machinery, leaving Britain to import them on a grand scale until the outbreak of war in 1914.

An electric power supply has to be generated using some other form of energy. A tiny generator such as the prototype invented by Faraday, in which a copper disc was spun between the two poles of a magnet, can be hand-cranked. Larger generators can be powered by watermills and windmills, and they were from the earliest days of electric power. But really large generators that could supply current over a wide region needed very powerful sources of energy to set them going. A rush of water many thousand times more powerful than a watermill produces hydro-electric power, a means of generation widely employed very early on in mountainous regions of Europe and America.

In Britain hydro-electric power was initially confined to Scotland. There was only one obvious source of power to drive Britain's first large-scale electricity generators, and that was a steam engine burning coal as its fuel. In the late nineteenth century coal mining was a huge industry employing about one million men, who produced enough 'black gold' to fill the nation's hearths, keep industry running, and supply the gasworks, with plenty left over for export. There was no problem finding fuel for electric power, but from the earliest days of the industry the generators caused environmental problems.

As electricity became more and more available to the mass of the population, and its use in industry and transport grew rapidly, the claim was made that it was 'clean'. It was true that electric lighting was much cleaner than gas. Electric trams and trains did not belch soot like steam railways. A factory which ran all its machinery and its heating on electric power was much cleaner than the old-fashioned model that pumped smoke and steam from its chimneys. In fact, once wired up to electric power you could imagine that this wonderful new source of light and power had solved all the environmental problems of the nineteenth century.

The fallacy was soon revealed. Large power stations pumped sulphurous smoke into the atmosphere and a cheap and efficient supply of electricity required the creation of a national 'grid', a web of wires connecting power stations over the whole country and carrying current into rural areas. When the building of this began in the late 1920s there was an outcry. Across the cherished landscape of the Lake District and the genteel southern Downs would march huge pylons, holding high-voltage cables on crooked steel arms. Was electricity, it was asked, that necessary? Could the cables not go underground? Throughout rural Britain battles were fought and won, chiefly, by the electricity lobby. In towns, huge power stations supplying the grid were denounced as polluting eyesores: the building of Battersea Power Station in 1933–7 caused a storm of protest from the inhabitants of Chelsea. Electricity might be 'clean' in the home and the factory, but it was ugly and polluting in the wider environment.

These problems, which beset the electrification of Britain when it was being established, have never gone away. In fact, they have become more severe, for our dependence on electricity is now more or less absolute, while our ability to generate it has become ever more problematic. The danger of Britain's former total dependence on coal for its electricity supply was first brought home in the severe winter of 1947. Coal stocks at the power stations had been far too low when an arctic front brought six weeks of blizzards and sub-zero temperatures throughout most of Britain. Mines were cut off, trains were snowed in, the collier fleet from the north-east was frozen in harbour. With no coal for the power stations, the lights went out and factories closed down.

Yet how would Britain generate its electricity in the post-war years? For a time the answer was 'nuclear'. But, as always, there were political considerations. If nobody had coal fires any more and no companies bought coal for their own steam engines and then the power stations went nuclear, what would happen to the miners? The same question arose when any alternative to coal – natural gas or oil – was suggested. Power stations worked just as effectively (though not necessarily as economically) whatever their fuel. In Scotland there was the possibility of hydro-electric power on a significant scale, but the scarring of the landscape beloved of the aristocracy for salmon fishing and shooting led to some fierce battles.

In 1956 the Clean Air Act was passed. Coal fires in the cities were outlawed and disappeared over the next decade. The last of the smogs that were such a feature of London winters descended in 1962–3. Urban households and factories were rapidly electrified, their chimneys no longer smoking in winter. To have gas lighting in the 1930s was not unusual, but to have it in the 1960s was archaic, even quaint. Town gas, manufactured from the coking of coal, was phased out and the great articulated holders mostly disappeared, the last few saved by preservation orders. Gas was piped in from the North Sea – 'natural gas' – still favoured for cookers and domestic water heaters and a cheap fuel for power stations.

It has been evident for many years that substituting gas for coal or oil for gas is not going to provide a solution to the generation of electricity far into the future. Yet every alternative scheme has its price: tidal power can only be harnessed by damaging the ecology of estuaries, wind power 'damages' the landscape, nuclear power carries dangers and so on. At every turn the key problem is not technology: that, it seems, is always there, ready with an answer. What gets in the way is politics, in the sense that social and economic considerations always divide opinion and can lead to a rejection of technologically feasible solutions.

Britain's electricity industry was run from the outset in the late nineteenth century as a mix of private enterprise and public ownership and control. Many local authorities were the proud providers of electricity for their ratepayers, and Parliament controlled the development of the industry with tight legal reins. In the period between the world wars, when electricity was made much more widely available, the development of the industry was overseen by a government-appointed Electricity Board. But it was not nationalized nor ever fully privatized. In 1948, the Labour government finally brought it lock, stock and barrel into public ownership, taking away from local government one of its cherished responsibilities. Then, between 1990 and 1996, the industry was entirely *privatized* for the first time in its history. At the same time the technology had developed so that Britain could exchange electric power across the Channel: the first cable allowing for this was laid in 1961. As an industry spokesman put it: 'We might now boil an egg in London with electricity generated in the Alps.'[2]

In fact, we have no idea where the electricity that we switch on in our homes and offices has been generated, except in the general sense that it has come from a power station connected to a grid which is connected to our homes. And we know that without that power we are in serious trouble. Electricity has insinuated itself into our lives in a surreptitious manner, adding this, and then that, little gadget in the home and taking over the powering of a large part of our transport system as well as commerce and industry. The promises and problems that electricity has presented us with since the first of the arc lights lit up town squares and public buildings have not really changed; only the scale of the issues is today vastly different from that of the late nineteenth century.

Were they alive today, the founders of the industry – Crompton, Swan, Ferranti, Siemens – would be astonished to discover that electricity generation is now implicated as a major cause of the over-heating of the planet. They believed they were ridding the world of the pungent pollution in which they were brought up, the air thick with coal smoke, their homes lit with candles, naked gas flames and oil, their roads covered in the dung of the horses that hauled the wagons, buses and trams of the cities. Electricity did bring in the cleaner, faster, brighter, more efficient world the pioneers envisaged, but at a cost that has yet to be assessed. It all began as a novelty and one which few thought would have much impact on the nineteenth-century world. In the twenty-first century, however, our total dependence on electricity is a major and very urgent political issue.

PART ONE

VICTORIAN DAWN
1870–1918

CHAPTER ONE

UNDER THE ARC LIGHTS

> It is somewhat remarkable that the first great triumph of the electric light should have been gained in the field of sport. On Tuesday last a football match was played at Sheffield by electric light ... At each corner of the ground set apart for play was a lofty structure – something like a huge inverted coal-scuttle, covered with a tarpaulin. There were also at each end of the field two small engines for generating the light, which was so dazzling as almost to blind the spectators when looking at it direct.

This report of the world's first football match played under floodlights appeared in the *Sporting Gazette* of 19 October 1878. The illuminated match was not by any means 'the first great triumph of electric light', but it did attract a great deal of interest and press comment at the time and inspired the *Gazette* reporter to wax lyrical about the life-enhancing possibilities of this new and intensely bright illumination.

> To the rich this will seem merely an opportunity of gratifying a fresh freak – a novelty, the zest of which will vanish like that of all other novelties. But to the hard-working professional man and to the 'toil-worn artisan' the thing will be more than a novelty – it will materially enlarge their sphere of recreation. Each will be able to enjoy the sport he delights in – racing, athletics, cricket, football – to the full, long after the sun has set. There will be no period to his recreation as there now is. He can come home from his work, enjoy his dinner comfortably, and then sally out, long after the shades of night have fallen, to the scene of his sport.

The football game was staged by an enterprising local businessman, John Tasker, not to advertise soccer, but to show what could be done with these brand new 'arc lights'.[1] He had installed a little generating station at his works in Sheffield so that he could replace gas lamps with electric arc lights and he considered the experiment a success. An enthusiast for new technology – he created Sheffield's first telephone exchange – Tasker believed that the football match might convince other factory owners to convert to arc lighting and that he could then supply them with the necessary equipment. He was not yet, however, a manufacturer of electrical equipment and had to hire what he needed for the match.

In 1878 Sheffield's Bramall Lane sports ground was, as a local reporter put it, a 'black wilderness' at night. All the events, both cricket and football, were played in daylight and there was no need for gas lighting in the ground. Like all industrial towns then, Sheffield had street gas lighting, but none of this penetrated Bramall Lane, which therefore provided Tasker with the ideal setting for his bold demonstration of the power of the arc light. He would need just four of these to turn the dark of an October night into daylight. The football match itself was staged especially for the occasion, with two brothers appointed as opposing captains who chose their teams from some of the better-known players of the day. It was played on a Monday, with a kick-off scheduled for 7.30 p.m. A very large crowd was expected and the police drafted in an extra seventy-five men in case of trouble.[2]

Behind each goal Tasker had placed a portable steam engine. Each engine was harnessed to two dynamos, which, when spun into action, created the power for the arc lights set on towers thirty feet high at the corner of the pitch. There was considerable difficulty getting the lamps to line up so that the light they threw covered the whole pitch, but after some rehearsals it was thought to be satisfactory. Reports on the game differ a little, some saying it was a triumph, others that on occasion the players became dazzled and confused. One intriguing observation was that women arrived with umbrellas which they opened to keep the bright light off their faces. Women in Paris had done the same when caught in the beam of the arc lights in the Avenue de l'Opéra. There was certainly no need for umbrellas otherwise at Bramall Lane, for it was a fine night and, in fact, moonlit. It was generally agreed that Tasker's lights put the moon in the shade. However, as far as floodlit football and Tasker's bid to go into the electric lighting business were concerned, this was a false dawn. Tasker did not pursue his interest in the arc light, and there was no floodlighting at the ground again until 1953.

In 1878 arc lighting was still quite novel, but there were a handful of manufacturers who could have lit Bramall Lane: Brush, Crompton, Jablochkoff were three of them. Tasker chose the generators and lamps made by the Anglo-German firm of Siemens, by then a well-established manufacturer of telegraph cables, with a sizeable factory in south London. The brothers Werner and William Siemens were from a large family born to farmers in Germany. Both proved to be brilliant engineers, with Werner, the elder brother, taking the initiative. In 1843, William was sent to England to sell a patent for a method of electroplating metals.

One of his favourite stories was that he was so ignorant of the language he mistook an undertaker's parlour for the patent office. In time, William settled in England, remaining in touch with Werner and working on new projects as the company's London agent. He took British citizenship in 1859 and in the same year married Anne Gordon, the daughter of Joseph Gordon, Professor of Electrical Engineering at the University of Glasgow. Not long before his death in 1884 he was knighted.

An interest in arc lighting came late in the career of William Siemens, for it was not until the 1870s that the technological problem of turning experimental lights into commercially available models was solved. How that came about can be told briefly, though it involved many years of determination and ingenuity, as the properties of electric current and its effects were discovered largely by trial and error.

To begin with there was the battery made by the Italian Alessandro Volta, which was the first really useful device for artificially creating a continuous electric current. The instructions for making his 'Voltalic Pile', an electrochemical cell on which modern batteries are based, were described in 1800 by the inventor himself in the *Philosophical Transactions of the Royal Society*. Humphry Davy experimented with this new gadget and demonstrated that a spark was produced if electricity from a large bank of batteries was passed over a small gap between two pieces of carbon. The miniature lightning flash was the 'arc' of the carbon-arc lamp.

Though Davy had demonstrated the arc light as early as 1808 the difficulties involved in making it practical took half a century to resolve. Carbon rods which burned slowly had to be made relatively cheaply. Mechanisms had to be devised to keep the rods the right distance apart as they burned away. And they burned unevenly, the 'positive' carbon rod disappearing faster than its 'negative' counterpart. The batteries based on Volta's design contained zinc and silver, and were therefore expensive, and yet they produced a feeble current unless many were used together. Nevertheless, a great variety of arc lights powered by batteries were made. Two Englishmen, W. E. Staite and William Petrie, tried to devise a serviceable battery-powered arc light between 1847 and 1851 and had some success. They lit up the portico of the National Gallery in 1848 and Staite toured the country giving demonstrations of his lamps. Though he failed to produce a commercial arc light, Staite's lectures, cut short by his death in 1854, were an inspiration to later inventors.

Electricity in a variety of manifestations was commonplace by the mid-nineteenth century, with the telegraph rapidly extending across the country and the newspapers humming with advertisements for such novelties as electric corsets and all kinds of bogus electro-medicinal cures. None of these devices required much in the way of electric current and batteries were adequate for nearly all of them. But if electric lighting was to have a future a new source of power would be needed which could supply a strong current over long periods of time.

The breakthrough had come in the 1820s when Michael Faraday, assistant to Humphry Davy, was experimenting with the relationship between magnetism and electricity. Faraday as a boy had been apprenticed to a bookbinder, but his interest in science had led him to the lectures given by Davy at the Royal Institution in London's Albemarle Street. Davy was impressed when Faraday sent him notes made at the lectures and took him on as a researcher. Battery power enabled those investigating the nature of electricity to conduct a number of experiments that illustrated its often surprising characteristics. For example, passing an electric current around an iron bar could turn it into a magnet. This suggested that reversing the process might *create* an electric current. Faraday spent a great deal of time on this proposition without success. He discovered, however, that if a magnet *moved* in relation to a coil of wire there was an effect. Crude devices were made which were hand-cranked. These were the first electrical generators, and Faraday demonstrated his prototype in 1831.

An infinite variety of generators was experimented with after this discovery. Although Faraday remained for most of his life a purely experimental scientist rather than a would-be inventor, when he was in his late sixties he was involved in one project to which his discovery had given rise. A very large, two-ton generator built by a Professor F. H. Holmes was used to power an arc light which was installed in the South Foreland lighthouse. Faraday oversaw the installation of the generator, driven by a steam engine, which first beamed light from the white cliffs of Dover in 1858.

The French began to install generator-driven arc lamps in their lighthouses in 1863 and were in the vanguard of developing new mechanisms for the lamps themselves. A Parisian inventor, Victor Serrin, devised an elaborate geared mechanism for holding the carbons in place as they burned away, and his design was used in a number of places in England. It seems that almost anyone who put their mind to the problem of how to improve equipment had a chance of success. Zenobe Theophile Gramme, a barely educated carpenter working in Paris, designed a generator that for a while was very widely used and had a real influence on the development of electricity in Britain. The son of a tax clerk, Gramme was born in Belgium in 1826 and began work as a joiner when he was still a boy. He later travelled in France, where, settling in the Paris suburb of Neuilly-sur-Seine when he married, he got work as a model maker for a French company manufacturing electric generators. Drawing on the work of others, Gramme in time produced a very efficient generator that was suited to powering arc lamps.

For a while the French were leaders in electric lighting. When the Prussians besieged Paris between September 1870 and January 1871, arc lights were stationed around the city's fortifications to discourage night-time assaults on the vulnerable parts of the defences.[3] All but one of these arc lights was powered by batteries. The single most powerful lamp, which could illuminate a large part of north-west Paris, was powered by a generator installed on the hill of Montmartre.[4] When the siege was over, peacetime Paris became known as the City of Light when arc lamps were switched on along some of the grand boulevards, notably Avenue de l'Opéra. Gramme generators were popular when harnessed to the ingenious arc lights invented by the Russian Paul Jablochkoff. In 1875 Jablochkoff had resigned his post as director of telegraphs between Moscow and Kursk and set off for America, with the intention of visiting the International Exhibition being staged in Philadelphia. But he did not get further than Paris, where he met a Frenchman, Louis Bréguet, inventor of electric clocks. Jablochkoff developed what became known as his electric 'candle', working in Bréguet's laboratory. Marketed by the Société Générale d'Électricité, Jablochkoff candles were for a time the state of the art in electric lighting.

By 1878, when the floodlit football match at Bramall Lane was staged, there had been a great many demonstrations of the arc light. Most of them, however, had been in Paris and it was a lament of British engineers that London and the other big cities were being left behind. One enterprising local authority in London, the Chelsea Vestry, despatched an engineer to Paris to report on the electric lighting there with a view to experimenting with it as an alternative to the existing gas lamps. By the time of his visit early in 1878, the Jablochkoff candles which lit up the Avenue de l'Opéra had been modified so that when one set of carbons burned out another set came on, giving a continuous light. With other models, the changing of the carbons in arc lamps could be costly and irksome. G. H. Stayton, Chelsea's engineer, was impressed:

> ... the light is vastly superior to gas, and is not injurious; there is an absence of the noxious smells both in the production and combustion; the heat in a room, so often unbearable in the case of gas, is scarcely felt; the most delicate colours are preserved; the air is not consumed as in the case of gas; there is no chance whatever of explosion, and, although the light is so powerful in the streets no accidents to horses have occurred.[5]

Despite this glowing account, Stayton thought it was too early to go electric in Chelsea. The lighting of the Avenue de l'Opéra was very expensive, with one Gramme generator to sixteen lamps. In fact, the Jablochkoff candles were switched off at midnight and 400 gas lamps brought on until dawn. Gas, despite its evident disadvantages, was cheap in Britain, as it was manufactured from the country's abundant supply of coal. Arc lighting, therefore, needed to find a niche in a lighting market dominated by gas. One of these was the lighting of industrial works, which were kept in operation well after dark and required a brightness that was difficult to achieve with gas. It was just such a project that brought into the infant electric lighting industry one of its most colourful and enterprising characters, the lavishly named Rookes Evelyn Bell Crompton.

Crompton was born in 1845 into a wealthy family which had a large estate in Yorkshire. At the age of six he was taken to the Great Exhibition in Hyde Park, where the Machinery Hall fascinated him. He began his education in a small private school where a fellow pupil was Charles Dodgson, who under the nom-de-plume of Lewis Carroll became famous as the author of *Alice in Wonderland*. Yet even that hugely popular Victorian children's fantasy hardly compared with Crompton's real-life adventures as a boy.

When he was ten years old, the 2nd West Yorkshire Light Infantry Militia was despatched to Gibraltar to relieve troops fighting Russia in the Crimea, a war which had broken out in the autumn of 1853. His father was an officer in the Militia and volunteered to lead the troops to the Mediterranean stronghold. As he was likely to be away for some time, he took his wife, daughter and younger son with him.

From Gibraltar young Crompton explored southern Spain with his father, riding for miles through wild country where they were shot at by bandits, surviving the attacks to visit Seville and Granada. Back on the Rock of Gibraltar he found that the naval ship HMS *Dragon*, which was commanded by Captain Houston Stewart, a cousin of his mother, was in port en route to the Crimea. In his *Reminiscences* he says he is not sure why his parents put him aboard the *Dragon*, but he was soon heading towards the battlefront, enrolled as a cadet in the Royal Navy. When he arrived in the Crimea he was given leave to visit an older brother in the trenches at Sevastopol. Here, in the last days of the war, he came under fire, an experience which entitled him to the Crimea Medal and the Sevastopol clasp. He was not yet twelve years old.

In 1856 Crompton went back to school, but he did not have much of an education: two years were spent preparing for Harrow, where he stayed for just two years. By the time he left Harrow in 1860 at the age of fifteen he had started to build a steam-powered road vehicle on his father's estate in Yorkshire and showed a great interest in engineering. He did not go to university but stayed around the family estate, working sometimes with his older brother. The family were not sure what he should do, and he was sent to Paris for six months, where he mingled in the most elevated company and learned some French, which turned out to be very useful when he first became interested in electric lighting. Finally a decision was made: he would go into the army. He took the necessary examination and at the age of nineteen joined the 3rd Battalion of the Rifle Brigade as an ensign. For four years he served in India, where he finished building his steam engine after it had been shipped out with all his tools. In time he managed to replace many of the Indian bullock carts with steam road engines he designed himself.

When he retired from the army in 1876, Crompton found work as an engineer, going into partnership with Dennis & Company of Chelmsford, Essex, a firm making horticultural equipment. He became a travelling salesman for the firm, always on the lookout for more work as he toured the country. It was his next move which got him into the lighting business. Relatives with an ironworks at Ilkeston in Derbyshire took Crompton on to supervise a new method of casting pipes, which involved a very large initial investment. They wanted to keep their Stanton Works going continuously with three shifts of workmen.

Crompton recalled in his *Reminiscences* the trouble they had with night-work: 'We tried portable lamps and various methods of gas lighting, but all were a failure. I had recently read of the developments of the electric arc light in France by M. Gramme, who was using a new form of electrical generator, or as it then began to be called, a dynamo-machine.'

Crompton got some Gramme generators and Serrin arc lamps sent over from France and lit the Stanton works electrically. It was a huge success, attracting a great deal of interest. With a partner from the Chelmsford works, a Mr Fawkes, who also spoke French, he formed a company importing and selling Gramme equipment. With one set he lit part of the newly opened London department store William Whiteleys in Westbourne Grove, near his home in Porchester Gardens. He would take guests to see the lights in the hope of getting more business. In his *Reminiscences* Crompton recalled that although he was a novice when it came to this new form of lighting, when he went to Paris with a party from the Institution of Mechanical Engineers in September 1878 he found himself 'regarded as a leading authority on electric lighting'.

Before the year was out he had formed his own electric lighting firm, Crompton & Company, and was intent on making improvements to the generators and lamps he had been importing. He took over part of the Chelmsford works of Dennis & Company and handed over the day-to-day running to a Swedish engineer, Andreas Lundberg, who had settled in England in 1862. Lundberg had begun his career as a mathematical instrument maker in Stockholm, before travelling across Europe working as a mechanic and engineer, for some while with the Russian Navy. He joined the Siemens brothers when they were laying the first Atlantic telegraph cables. When Crompton hired him, Lundberg was running his own business and took the job on the understanding that it would last only as long as it made a profit. Under Lundberg's management the firm proved successful, selling two different kinds of lighting equipment. In one a steam road vehicle hauled the generators and then powered them when they were set up for lighting;

in another the steam engine and generators were mounted on to horse-drawn carts.

In his *Reminiscences* Crompton wrote: 'With this first portable set of electric light arc generating plant and several others of the same pattern we carried on a regular advertising campaign, principally in the Home Counties, besides hiring out sets to contractors for public works, so they could carry on their work at night.' He and his engineering assistants 'became scientific travelling showmen' as they adventured across the country. In July 1879 he took his portable equipment to Henley Royal Regatta and lit the fashionable gathering at night. It was a success, he recalled, but attracted the unwanted attention of some young blades who thought it would be a fine prank to plunge everyone in darkness by cutting the cables. They reckoned without the valour of the Crimean veteran and his engineering crew, which included Peter Willans, who made his steam engines. In the ensuing battle Crompton recalled: 'Spanners were freely used, and it was rumoured that certain members of the Jesus boat were not able to row on the following day.'

It was in 1878 that British lighting engineers found they were in competition with some potentially powerful rivals in America. The patents of Charles F. Brush of Cleveland, Ohio, were being touted in Britain and plans were afoot to set up the Anglo-American Brush Electric Light Corporation, which was finally incorporated in 1879. Brush was a chemist who went into business with a friend who ran a company supplying equipment for the electric telegraph. Working alone, he first designed a very efficient dynamo and then made improvements to the arc lamps then available, paying special attention to the quality of the carbon rods used. He lit the centre of Cleveland in 1879 and supplied a commercial company in San Franciso with arc lighting equipment in the same year, an undertaking which resulted in what some say was the first central power station in the world. Brush was able to sell his equipment and licences to produce it all over America in the brief heyday of arc lighting, and he became a wealthy man. He built himself a mansion in 1889, in the grounds of which he built a giant windmill, which turned the dynamos that supplied him with electricity, some of it stored in batteries in the basement of the mansion. For a while the name Brush was almost synonymous with electric lighting in Britain, as many companies were formed to promote his technology. However, he was to be quickly overshadowed by an American rival.

On 8 October 1878 the London *Times* reproduced from the 16 September edition of the *New York Sun* an interview with the lionized inventor Thomas Alva Edison. Earlier that year Edison had astonished the world with his 'phonograph', which could record and play back speech and music. He had chanced upon the discovery while trying to make a kind of answer-machine for the telephone, which was then still a novelty. The phonograph was entirely mechanical and a commercial failure but it gave Edison the title 'The Wizard of Menlo Park', after his rustic research station in New Jersey.

In an interview with Edison the *New York Sun* reported: 'Mr Edison says that he has discovered how to make electricity a cheap and practicable substitute for illuminating gas. Many scientific men have worked assiduously in that direction, but with little success. A powerful electric light was the result of these experiments, but the problem of its division into many small lights was a puzzler. Gramme, Siemens, Brush, Wallace and others have produced at most ten lights from a single machine, but a single one of them [arc lights] was found to be impracticable for lighting aught save large foundaries [*sic*], mills and workshops. It has been reserved for Mr Edison to solve the difficult problem. This he says he has done within a few days.'[6]

Edison was a latecomer to electric lighting, taking an interest only in September 1878 when he visited a firm making generators and arc lights. His vision of an electrically lit world was instant, and in a mood of great excitement he returned to Menlo Park and began to put together a team of researchers and engineers to work on the problem. One of these was the young physicist and mathematician Francis Upton, who had graduated from Princeton and had studied in Berlin. Edison, on his own admission, was hopeless at mathematics and had no science training at all. His expertise was in thinking up things to invent and putting together a team, and the funds, to carry it all through. Upton was set to work checking out all existing patents on electric lighting, while Edison excited the newspapers with his rather exaggerated claims for his discoveries.

He told the *New York Sun*: 'When the brilliancy and cheapness of the lights are made known to the public – which will be in a few weeks, or just as soon as I thoroughly protect the process – illumination by carburetted hydrogen gas will be discarded. With 15 or 20 of these dynamo-electric machines... I can light the entire lower part of New York City using a 500 horse power engine.' Edison then outlined all the technicalities of meters and cables. He would use existing gas lamps in the street for his public lighting, and gas fittings in the home for electricity. 'Again, the same wire that brings the light to you will also bring power and heat. With the power you can run an elevator, a sewing machine, or any other mechanical contrivance that requires a motor, and by means of the heat you may cook your food.'[7]

When in 1879 Edison applied for a British patent for his lighting system, it became clear that he did not have in mind the arc lights which were being switched on in many parts of the country. What he was working frantically to make practical was a new kind of light entirely, in which a wire of high resistance material heated by a strong electric current would glow inside a glass container. This had been the Holy Grail of inventors for many years, but they had nearly all been confounded by the practical difficulties that Edison was boasting he had solved more or less instantly.

He and others were not far off, but in the meantime the limitations of the arc light were becoming clear.

Arc lighting had been a great success in railway stations and in large buildings and open spaces. Crompton had lit St Enoch's station in Glasgow, and in London the Metropolitan Board of Works, responsible for the building of the new sewers and the Embankment, had lined the Thameside pavements and some bridges with Jablochkoff candles, supplied and operated by the French Société Générale d'Électricité. A round of applause had greeted the switching on of arc lights provided by Siemens in the British Museum Library reading room in 1878. Until then it had had to close when darkness fell – which was quite often during the day in foggy weather – because it was too damaging to the books to have gas light.[8]

It was beyond dispute that the electric arc light was not as polluting as the gas lamp, which still then, in 1878, burned with a naked flame. But its brightness was a real problem. In a letter to *The Times* of 30 October 1878, C. Meymott Tidy, Professor of Chemical and Forensic Medicine at the London Hospital and Medical Officer of Health for Islington, wrote: '"Gas v. Electric Light" is a cause just now being tried before a public tribunal. The questions of vested interests, value of gas shares, and the like, fortunately do not concern me. Allow me, therefore, as an independent witness, to say a few words in the cause under discussion.' After a discourse on the advantages of gas, chiefly because its brightness could be easily adjusted, the doctor recounted his own experience with the arc light, which was not encouraging:

> I have made a point of remaining in the neighbourhood of the electric light for at least three hours in order to observe its effects upon me... before long my eyes became entirely blinded to all rays except the blue; and, as a result, everybody and everything appeared of a ghastly blue tint. For hours after I returned home, the blue rays haunted me; but what was worse still, I suffered from what I am rarely a sufferer from – an intense headache, especially seated about the region of the eyes. As a medical man, I am convinced that whatever may be the advantage of the electric light as an illuminant for large outdoor spaces, it can never be used as a room

> illuminant, or even as a general street illuminant with advantage, save to the medical profession generally and to ophthalmic surgeons in particular.

Nonetheless, there were arc lights fitted out in some grand homes. William Armstrong, the hydraulic engineer and arms manufacturer, had one in the picture gallery of Cragside, his wild country seat in Northumberland, the current provided by a hydro-electric generator. And Lord Salisbury upset his family by having arc lights indoors at Hatfield House, though he was soon told to get rid of them. His Jablochkoff candles on one occasion set fire to the curtains in the dining room and were not appreciated by the ladies. His daughter recalled in a family memoir: 'For a brief period family and guests were compelled to eat their dinners under the vibrating glare of one of these lamps in the centre of the dining hall ceiling.'[9]

By 1879 there was such a hubbub of interest in electric lighting and so many conflicting views that a House of Commons Select Committee was appointed to investigate, chaired by the distinguished chemist Lyon Playfair.[10] The brief was simply to inquire into the issue of whether any new laws were needed to determine how electric lighting schemes might go ahead. A great deal of evidence was collected, including a report on how a scheme had been received at Billingsgate Fish Market. This was a project of the Société Générale d'Électricité, whose representative explained to the committee that the salesman in Billingsgate missed the warmth from the gas and did not like the glare of electric light on the fish, partly because it might put off lady customers. A report by the London correspondent of the *Liverpool Mercury* on the experiment offered a rounded judgement on the value of electric light at the time:

> Though diamond merchants declare that they can sell diamonds by the electric light and convey the impression that they can also buy them; though compositors are said to like the new light; though

> when used at the Royal Institution it was found an admirable substitute for gas; yet the fishmongers declare that it will not suit them. It may do for Albemarle-Street, but not for Billingsgate. The trial was made among the fishmongers yesterday. When the light streamed over the dark and murky Thames, it was at once recognised as better than gas for outdoor work; but the fish salesmen declared that they did not know herring from mackerel under the new conditions. As the Billingsgate trade is mainly conducted in the dark, this declaration is somewhat disheartening to the people who wish to rid themselves of the gas companies and all their works.[11]

The fishmongers got their gas back, and the brief two-page conclusion of the Playfair Committee in 1879 was that it was too early to say who should run electric lighting schemes in future: local authorities or commercial companies. The technology was changing too rapidly. And, sure enough, within a couple of years everything everyone knew about electricity was out of date. Edison's vision of thousands of twinkling little electric lamps lighting up New York and other great cities seemed as if it might soon become a reality.

CHAPTER 2

SWAN SHOWS THE WAY

One of the companies Crompton had supplied with generators and arc lights was the firm of Mawson & Swan, chemists in Newcastle upon Tyne. In the late summer of 1880 a representative of the firm called on Crompton at his London office in Queen Victoria Street, with an urgent request that he leave with him immediately and take the first train north they could catch. There was no time to go home to get any clothes and Crompton would need to telegraph his wife to say he would not be home that evening. 'I accordingly went to Newcastle,' wrote Crompton in his *Reminiscences*, 'where Swan took me into his laboratory and showed me twenty small incandescent lamps, which burned very brightly and steadily, each having a carbon filament enclosed in a globe, exhausted to a very perfect vacuum. He claimed, and I agreed with him, that he had solved the problem of electric light for internal illumination.'

The man Crompton met, Joseph Wilson Swan, did not have the

international reputation of Thomas Edison, nor was he a keen self-publicist like the American. A cautious, methodical worker, Swan was fifty-two years old when he summoned Crompton to observe his newly fashioned light bulb. He had been working on this project on and off for more than twenty years and had made a crude kind of incandescent light bulb as early as 1860, when his rival Edison was just thirteen years old.

The familiar domestic light bulb, with its bayonet or screw fitting, in pearl or plain, offering 40, 60 or 100 watts, took many years to create. A great deal to do with electricity is counter-intuitive, and the fact that this new form of illumination was the first electrical appliance that required really large-scale power stations for its commercial success is perhaps surprising. But heating the flimsy filament inside the bulb so that it becomes white hot requires a very strong current. Until there was a generator that would do the job, experimentation with different materials for the filament was not really practical. It was understood, too, that the filament had to burn in a vacuum if the glass bulb was not to blacken, but there was nothing available to extract air from a glass bowl in 1860.

The story of how Joseph Swan first became interested in electric lighting and finally achieved success with his first saleable incandescent lamps in 1880 does not cover the whole of the long process of invention, but it gives some idea of the painstaking efforts of the many people involved.

Swan was born in the north-eastern coastal town of Sunderland in 1828, at a time when gas lighting was still a novelty. It was used chiefly in factories and to light the streets at first, and it was rarely found in the homes of the north-east of England. Domestic lighting had really not changed much in several centuries for most of the families in Sunderland and the surrounding countryside, as Swan recalled in an account he gave of his childhood:

> The days of my youth extend backward to the dark ages, for I was born when the rush-light, the tallow-dip or the solitary blaze of the hearth were common means of indoor lighting, and an infrequent glass bowl, raised 8 or 10 feet on a wooden post, and containing a cup full of evil-smelling train-oil with a crude cotton wick stuck in it, served to make the darkness visible out of doors. In the chambers of the great, the wax candle or, exceptionally, a multiplicity of them, relieved the gloom on state occasions, but as a rule, the common people, wanting the inducement of indoor brightness such as we enjoy, went to bed soon after sunset.[1]

The only really significant improvement in lighting for the home by the time Swan was born was the lamp devised in 1781 by the Swiss chemist François-Pierre Ami Argand. His lamp had a round wick so that air was drawn up through the centre, causing the flame to burn more brightly. When a glass chimney was added, and then a globe around that, and then an adjustable wick and the whole placed on an ornate stand, Argand lamps became popular in the wealthier homes and remained in common use right through the nineteenth century. But they were beyond the means of most working-class families.

Even the cheapest tallow candles made from animal fats were regarded as something of a luxury by the poor. From 1709 until 1831, three years after Swan was born, candles were taxed and in the countryside the 'rush-light' was favoured as a cheaper alternative. In his *Cottage Economy*, published in 1822, William Cobbett, the radical and reformer, wrote: 'We were not permitted to make candles ourselves, and if we were, they ought seldom to be used in a labourer's family. I was bred and brought up mostly by rush-light, and I do not find that I see less well clearly than other people.'

The rush-light was still in use long after gas lighting was available and in her *Old West Surrey*, published in 1904, the gardening writer Gertrude Jekyll quoted a ninety-year-old who showed her how it was done. 'You peels away the rind,' she told Jekyll, 'and when the rushes is dry you dips 'em through the grease, keeping 'em well under. And my mother she always laid hers to dry in a bit of hollow bark. Mutton fat's the best: it dries hardest.' When cottagers went to bed, Jekyll wrote, they would 'lay a lighted rush-light on the edge of an oak chest or chest of drawers, leaving an inch over the edge. It would burn up to the oak and then go out. The edges of old furniture are often found burnt into shallow grooves from this practice.'

This was the world into which Swan was born. His father, John, was a ship's chandler selling provisions to the seafaring trade, a man who was disappointed not to go to sea himself and not very successful at his trade. He had a large family, four boys and four girls. Joseph had one elder brother, John, with whom he spent much of his time as a youngster exploring the streets of Sunderland, where he took an interest in all the trades, from cobbler to the candlestick maker. Though he had little formal education, the world in which he lived was a hive of activity and full of fascinating gadgetry. Joseph sometimes stayed with a great-uncle, Captain John Kirtley RN, an old sea-dog who had fought under Nelson. There, in the countryside near Sunderland, he would help out in the local carpenter's shop and the blacksmith's and with the cutting of the hay at harvest time.

For a while Joseph and his brother went to a dame school, of which there were a great many in nineteenth-century England before education became compulsory. The fees were low and the schooling basic, but here Joseph learned to read and write and to sew, a skill useful not only for future sailors but for the delicate work required for some of his later experiments. One event that created a strong impression was when a glass prism was shown to them and cast a rainbow on the wall of the school when the sun shone through it. Electricity was not to be seen much, but a family friend had a collection of batteries and 'machines' with which he could create special effects that might literally make your hair stand on end.

With his older brother, John, Joseph spent two years in another 'higher grade' dame school where a book on chemical experiments interested him. A family friend was a druggist who allowed him the materials to make gunpowder and other explosives on a small scale. His schooling came to an end when he was just thirteen, as his father could not afford the fees and the family was in financial trouble. He was lucky to find an apprenticeship with a firm of chemists in Sunderland. He had signed in 1842 to serve for six years before he was free from his articles, but both partners in the firm of Hudson & Osbaldiston died before three years were up, leaving Joseph free to move on at the age of seventeen. He joined a friend, John Mawson, who had established a chemist's shop, and began as a partner in business for the first time.

Swan continued his education with membership of the Sunderland Athenaeum, which was opened in 1841 as place where the town's intelligentsia could keep up to date with magazines and books and attend lectures by some of the leading scientists and political activists of the day. It was here, in 1845, that Swan first came across a description of a kind of incandescent light bulb. The Athenaeum library had a copy of the *Repertory of Patent Inventions*, which listed for that year a patent in England for a kind of electric lamp in which a carbon pencil enclosed in a glass vacuum was made to glow. The patent was in the name of E. A. King, who was a lawyer, and not that of the inventor, who turned out to be a young American called John Wellington Starr. Not a great deal is known about Starr, and the few stories of his short life are not consistent, but it seems he came to England with an electric lamp invention around 1845, exhibited it in one or two places, obtained a patent and then, quite suddenly, died. According to one account he is buried in Birmingham.

Swan did not get a chance to get to see Starr or his lamp, but there were others working on the same idea who did lecture at the Sunderland Athenaeum. One of these was William Edward Staite, the inventor and enthusiast for electric light well ahead of his time. Staite travelled the country demonstrating his 'regulator lamp', which was an early version of an arc lamp in which the burning of the carbons was kept even by a clockwork mechanism. Swan saw him several times when Staite came to the Sunderland Athenaeum in the 1840s and was able to take a close look at all the equipment he brought to his lectures. He also heard him speculate on the possibility of a new kind of lamp that might be made with a piece of platinum wire.

This inspired Swan to experiment with different materials which might form some kind of thin carbon filament that could be made to glow brightly when a current was passed through it. He managed after many experiments to produce carbonized paper which was thin but durable. He then attempted to 'light' this in a glass bottle from which most of the air had been excluded, the partial vacuum sealed with a rubber stopper. With a bank of batteries he did get a carbon strip to glow but it soon burned out. He continued to experiment until 1860, when he abandoned the project, as there seemed no hope at the time of making anything practical and commercial.

Staite died in 1854, still in his mid-forties and before arc lamps similar to his were much used. For Staite, Starr and Snow, and all the other early pioneers of electric lighting, the only source of electricity was a battery. That greatly limited the power they could generate and made it prohibitively expensive. Real advances with electric light were not possible until the first efficient generators became available. This was why Swan did not pursue his interest in lighting and concentrated instead on the chemical-electrical techniques of 'electrotyping' and photography. He invented and patented a number of chemical processes for fixing photographic images.

Swan and his partner John Mawson had a prosperous business. Swan married a teacher, Frances White, in 1862 and settled into family life in Newcastle upon Tyne. Then tragedy struck twice within a year. His partner, then Sheriff of Newcastle, took charge of a dangerous job disposing of a large quantity of the explosive nitro-glycerine which had been found abandoned in some stables. It was decided to cart it out to the wilds of Town Moor, to be buried in a deep gully. On the way it blew up, killing Mawson and all the men he had engaged to transport it. That was in 1867. The following year, Swan's wife, and by then the mother of three surviving children, died. Later Swan married her sister Hannah, the wedding taking place in Switzerland as it was illegal in Britain to marry a deceased wife's sister.

Swan never lost his interest in the possibility of electric light and by the mid-1870s there had been technological advances which promised to solve several problems that had thwarted him in his earlier experiments. First there were available the first efficient electricity generators, which were being used for arc lighting. It was possible to pick and choose: a Gramme or a Siemens or a Crompton. Then the German chemist Hermann Sprengel invented a simple but effective device for creating a vacuum in a glass container. This was crucial, as it was known that a carbon filament would burn out quickly and blacken the glass if air was not excluded.

The Sprengel pump was used by the scientist William Crookes to create what he called a 'radiometer', in which little squares of metal would turn on a pivot enclosed in a glass vacuum when sunlight fell on it. These are still made as 'executive toys'. When Crookes exhibited this little device in 1875, Swan's interest in electric lighting was rekindled. His eye was caught by a newspaper advertisement placed by a young bank clerk, Charles H. Stearn, who had a special interest in Crookes's radiometers and was experimenting with vacuums. Stearn was based in Birkenhead on Merseyside, and Swan wrote to him to ask if he could enclose one of his carbonized paper filaments in a glass bulb.

Stearn took on the job with the help of Fred Topham, a skilled glass-blower who made the bulbs. In theory Swan's carbonized paper and other filaments should have glowed brightly in the evacuated bowls, but they burned up quickly and turned the glass black, the same problem that had arisen with his earliest attempt at making an incandescent lamp.

The temptation was to give up, but Swan, with his extensive knowledge of chemistry, guessed that something was happening to the vacuum as soon as the bulb was turned on. His solution, which Stearn and Topham managed to make practical, was to continue to draw air from the bulb while it was first illuminated. This worked, and Swan, with his collaborators, finally had a saleable incandescent light bulb. Stearn, it was said, wanted to patent it straight away, but Swan, even though he had read about Edison's experiments, thought it would be a waste of time. He believed the concept of the filament light bulb had such a long history it would be impossible to claim it as his own idea.

At first Swan's lamps were made by a kind of cottage industry involving his family (he had five children by his second wife, Hannah), the glass-blower Fred Topham and Stearn the banker. By 20 October 1880 he had sufficient bulbs to dazzle a meeting of the Literary and Philosophical Society of Newcastle upon Tyne at the conclusion of a lecture on the subject of 'Electric Lighting'. With a flourish of showmanship Swan asked for the seventy gas jets which lit the lecture theatre to be turned off, plunging the distinguished gathering into darkness. After a moment, the generator was turned on and twenty Swan incandescent lamps began to glow, filling the theatre with light in a manner that was regarded as almost miraculous.

Swan repeated the demonstration a month later before a gathering of the Society of Telegraphy Engineers, attracting the attention of many of its members, including an enterprising young man called Henry Edmunds, who had just returned from America where he had met Edison and several other inventors and engineers working on electric lighting. In a series of articles recalling this exciting time, Edmunds gave an account of his first meeting with Swan, when he first learned of the new lamp. 'I immediately telegraphed to Mr Swan for an appointment; and he courteously invited me to Newcastle, where I was surprised and delighted to see the showroom of Messrs Mawson & Swan illuminated by a number of glass bulbs.'[2]

These were not the first electric lights Edmunds had ever seen. Born in 1853, he was from Halifax in Yorkshire and had been educated privately before training as an engineer. His family were well connected and in 1877 the local Member for Parliament passed on to him an invitation to witness a demonstration of Paul Jablochkoff candles at the West India Docks in London. It was characteristic of the gung-ho Edmunds that he jumped straight on to a train to London and arrived at the docks in time for the first showing of the arc lights. This was a failure, as the generator broke down, but it was inspiration enough for Edmunds to have a go at making his own arc light when he got back to Halifax. When Jablochkoff had solved his technical problems, Edmunds went back to the Docks to witness a successful demonstration of the electric 'candles'. He met there a man called Richard Werdermann, who gave Edmunds his business card and invited him to lunch the next day in the Palmerston restaurant in the City.

Werdermann had his own arc light design, which he wanted to sell in America. Anticipating that the Jablochkoff demonstration in London would create interest in the newspapers, he asked Edmunds to sail for New York to promote his version of the arc light. The arrangement that Edmunds agreed to was that he would have to pay his own fare but he could share any sales 50–50. 'On my return home,' Edmunds recalled, 'I informed my father that I was going to America, and he immediately asked if I knew anyone there. I replied, No! but I had an acquaintance who had recently returned and I thought he might give me an introduction.'[3]

Over the next year Edmunds criss-crossed the Atlantic, a voyage that then would take ten to twelve days. He worked as a kind of go-between, taking British lighting equipment to America and bringing American equipment back. At Menlo Park he met Edison just after he had invented the phonograph, and he returned with the first model ever to be exhibited in England. Because he had first-hand knowledge of rival systems, especially that of Edison, Edmunds was of special interest to Swan. He remained friendly with both men, but it was Swan he represented, introducing his incandescent lamp to America and working as part of a team promoting the invention in Europe.

By February 1881, Swan was confident he could make his light bulbs on an industrial scale and set up his first company to manufacture them. He got financial backing chiefly from Newcastle investors and found a site for a factory in the Benwell area of Newcastle upon Tyne. Rookes Crompton became the chief engineer of the Swan Electric Light Company, and the glassblower from Birkenhead, Fred Topham, was brought in to teach the art of creating the little globes for the bulbs. At first, the only glassblowers Topham could find with the necessary skills had to be enticed over from Germany. Women were employed for the delicate work of mounting the filaments.

Once there was a supply of bulbs, Henry Edmunds was able to act as travelling salesman for the new enterprise. Swan could not pay him a salary but gave him a generous commission on every lamp he sold. However, take-up was sluggish at first, as the price was 25 shillings a lamp, around £60 at 2009 prices. One of his first assignments was to sell the lamps to a Royal Navy Captain, who had seen them at a dinner party in London in the spring of 1881 and thought they might do for his ship. Swan despatched Edmunds to Portsmouth; he remembered travelling 'with a small shallow box, containing lamps carefully packed in cotton wool'. The lamps, he recalled, looked similar to Crookes's radiometer and were quite large and cumbersome compared with the later models.[4]

The Captain, J. A. Fisher (later Admiral Lord Fisher), entertained Edmunds to lunch before arrangements were made to demonstrate the Swan lamps in a darkened shed in the Portsmouth dockyard. A searchlight generator had been set up outside the shed, with bare wires hung from strings providing the electrical link to the lamps. To get the generator running at the right speed, a bo'sun outside communicated by whistle to another inside the shed until the lamps glowed at the correct luminosity. At this point, Captain

Fisher arrived with a party of ladies. He asked Edmunds if he had seen the powder magazine in his ship RN *Inflexible*. What would happen to the lamps if his ship were hit broadside?

A sailor appeared with a tray on which there was gun cotton covered over with black gunpowder. Edmunds was asked if he was prepared to break one of the lighted lamps over the tray. He obliged, smashing 25 shillings' worth of glowing Swan bulb with a chisel. The light went out and a few bits of glass fell on the tray. There was no explosion or flare. Fisher considered for a moment, Edmunds recalled, and then announced: 'We'll have this light on the *Inflexible*.'

From Portsmouth Edmunds took a box of Swan bulbs to Bristol, where the Inman Steamship Company had asked to see a demonstration with a view to fitting out one of its liners, the *City of Richmond*, with electric lighting. A demonstration at the works of the tobacco company Wills, which had as a consulting electrical adviser the distinguished scientist Silvanus P. Thompson, was successful. More exciting for Edmunds was the decision to fit the *City of Richmond* with Swan lamps, and it became the first Atlantic liner to be lit with filament light bulbs. Edmunds had not long been married, and wrote to his young wife at their Yorkshire home to tell her of his triumph. But she did not see the letter or speak to him again, for when he returned to Halifax he learned that she had died a few hours after giving birth to twins, a boy and a girl, who survived.[5]

Joseph Swan invited Edmunds to Newcastle and advised him that there was nothing like work to 'relieve his great trouble'. Soon he was in Glasgow with more Swan lamps, this time taking an order for the *Servia*, the first of the Cunard liners to be fitted with electric lighting. In his 'Reminiscences' Edmunds recalled how primitive the installations were around 1881: 'Just imagine! We had no accessories whatever – they had not yet been designed. We had no sockets, even, for the lamps; we had no switches, no cut-outs (fuses), no measuring instruments. Everything had to be designed and planned for each fresh application.'

While Edmunds was Swan's travelling salesman, the inventor himself was involved fitting out his own house and those of some of his friends with the new light bulbs. The first installation in a private house, according to Swan himself, was the wonderful Victorian mock Tudor fantasy of Cragside, built for his friend Sir William Armstrong by the architect Norman Shaw. Though Armstrong was best known for making naval guns and other armaments, his first ventures in engineering involved hydraulic machinery. A native of the great mining city of Newcastle upon Tyne, Armstrong believed coal was used far too liberally and that sooner or later Britain would run out of it and would need to find an alternative source of energy. His Cragside estate, set in a kind of rocky moorland wilderness, was ideal for hydro-electric power. Armstrong had installed a Siemens generator, which was turned by a hydraulic pump of his own design. This lit the single arc lamp in his picture gallery before Swan installed the new incandescent lamps.

The arc lamp in the gallery was taken down and the generator wired up to the new light bulbs, with copper wires over a mile long carried on telegraph poles. 'It was a delightful experience for both of us when the gallery was first lit up,' Swan wrote later. 'The speed of the dynamo had not been quite rightly adjusted to produce the strength of current in the lamps that they required. The speed was too great and the current too strong. Consequently the lamps were far above their normal brightness, but the effect was splendid and never to be forgotten.'[6]

By the summer of 1881 Swan had made the manufacture and installation of incandescent light bulbs a going concern. They lit the Savoy Theatre in London and were being installed in Mansion House in the City. One or two local authorities were taking an interest in them as a possible substitute for gas lighting. Though there were many models of arc lamp on the market, Swan had only one rival in Britain for his incandescent lamp. This was St George Lane Fox, son of the famous explorer Augustus Lane Fox Pitt-Rivers. St George took an interest in electrical appliances as a teenager and developed his incandescent lamps independently, but was unable to produce them commercially at first. He sold on his patent to the Anglo-American Brush Light Corporation, which had been founded to provide arc lighting.

Swan was the leading manufacturer of light bulbs in his own country, but he was aware of impending competition from America. Not only was Thomas Edison intent on breaking into the British and European markets, there was another American, Hiram Maxim of the United States Electric Light Company, who had his own version of the bulb. All four inventors, it should be emphasized, were indebted to Hermann Sprengel, the German scientist working in London, for his beautifully simple and efficient 'pump' for extracting the air from the glass globes of the light bulb. It was a drip-feed mechanism, in which droplets of mercury carried away particles of air as they passed a tube fitted into the neck of the bulb. Sprengel first demonstrated his invention in 1865. Only when it became widely used to extract nearly all the air from glass bulbs could the search for the best material for the filament of the incandescent light begin.

In 1881, Maxim was a genuine rival to Edison, and, as a character, even more colourful, if less well known to the American and European publics. Born in 1840 in the backwoods of Maine, he had not much in the way of formal education. His father, descended from Huguenots, was a farmer and wood-turner and something of an inventor himself, toying with a design for an automatic gun and a flying machine. Maxim learned wood-turning and began a long series of jobs working for carriage-makers and travelling up into Canada, where he was for a time a bartender. When he was twenty-four he joined a firm of engineers owned by his mother's brother in Fitchburg, Massachusetts. From that time on he worked as an engineer, first for a gas company and from 1878 for the United States Electric Light Company.

Maxim had married an English woman in 1867 and fathered two children when his young wife discovered that he had taken a fifteen-year-old girl as a lover and that she had borne him a child. He returned to his first wife (it is not clear if he was technically a bigamist), only to leave her again when he had an affair with his secretary, with whom he travelled to Europe in 1881. Maxim's appearance on the electric lighting stage was brief: he settled in London in 1884 after selling his lighting patents to Edison and began manufacturing guns. When his quick-firing Maxim automatic gun was adopted by the army, he took British citizenship and was knighted in 1901.[7]

What drew Maxim to Europe initially was the world's first ever international exhibition of electrical appliances, which was staged in Paris from August to November 1881. He exhibited a method of coating carbon filaments and a regulator for electricity systems, as well as his own version of the incandescent light bulb. In fact all four light bulb makers, Maxim, Edison, Lane Fox and Swan, exhibited in Paris in a competitive atmosphere that one French electrical journal called 'The Battle of the Lamps'.

CHAPTER 3

LA BATAILLE DE LUMIÈRES

What the French called '*La Bataille de Lumières*' was, on the one hand, the rivalry between the makers of the first light bulbs and, on the other, the contest between electric light and gas light. In the view of the British technical journal the *Electrician*, it was the rivalry with gas that was the more significant, as there was not a great deal to choose between the lamps produced by Edison and Swan. Even those of Maxim and Lane Fox, while perhaps inferior, worked well enough. There were no embarrassing failures at the International Electrical Exhibition in Paris in 1881, though there was an attempt by Edison's representatives – he did not travel himself – to feed stories to the newspapers about the poor quality of the other displays.[1] But a report in the *Electrician* published on 22 October, towards the close of the Exhibition, made it clear that the incandescent light had not yet made the critical breakthrough that might make it a commercial success and a real rival to gas lighting.[2]

After listing all four makers of light bulbs, the report stated:

> We believe the incandescent electric light to have attained the point when it can successfully compete with gas, if it can be conclusively proved that either one of the above type of lamps has an average life of a thousand hours when used in consecutive periods of four hours, and can be sold to any purchaser for a florin [i.e. two shillings]. The representatives of Mr Edison state that the most recent figures give as the average life of his lamp something over a thousand hours, and that his system of manufacture has been so carefully considered that the lamps can be produced for a very small sum, less even than a florin. The evidence we require, however, is of a different kind – it is purely commercial. We ask where is the shop in London, Birmingham, Liverpool, Edinburgh etc. at which we can purchase the lamp, and what the price to the casual purchaser? When such shops exist we shall have the evidence required by practical men.

It was the boast of Edison's representatives, led by his loyal English assistant, Charles Batchelor, and the Menlo Park librarian, Otto Moses, that they were not just offering a light bulb for exhibition. They had 'a whole system' of generators and cables and meters and fittings of various kinds. Edison had ordered for the exhibition an especially powerful generator nicknamed the 'Jumbo', which was shipped across the Atlantic in 137 crates and arrived late, so that Batchelor and Moses had to 'talk up' their system before anyone saw it working. Once it was in place the Edison show was impressive. But there really was no way of finding out how much it would cost to operate.

The Swan Electric Light Company lacked the financial backing of Edison but put up a lively show through its enthusiastic representative and itinerant light bulb salesman Henry Edmunds, who recalled in his 'Reminiscences': 'I was young. Everything was fresh and new.

Paris was just recovering from the siege and the civil warfare of 1871. The Exhibition was in the Palace de l'Industrie in the Champs Elysées. All nations were assembled there. We had every known kind of apparatus connected with Telegraphy, Telephony, and Lighting, Arc Lighting, Incandescence Lighting, Dynamos and Electric motors of that period.'

The government refused to provide any financial help to Swan and the other British exhibitors, and left the appointment of official Commissioners to the last minute, sending the Earl of Crawford and Balcarres, whose elaborate title gave the French the impression that he was two people – or so one young wag claimed. The Earl was in fact a distinguished scientist who was to play a significant part in the creation of the largest power station in nineteenth-century Britain. However, he had no funds for Swan and his men, who had to rely on others. Edmunds had put together a team of men from his native Yorkshire to look after the machinery in Paris, and found funding for them and himself from a New York patent agent, T. J. Montgomery, who was keen to promote Swan lamps in America.

Edmunds would pick up his cheques from a Paris bank in lots of £1,000 or so. He recalled one occasion when he very nearly had to tell his team that he could not pay their wages. When he went to get his money the bank manager demanded some form of identification. There were no national passports then, and Edmunds was told that calling cards and letters addressed to him personally would not do. In desperation Edmunds offered to take off his shirt, which would have a tab with his name on it: he did not want to lose it in the hotel laundry. At this the bank manager relented and handed over the money. When Edmunds undressed that night he was astonished to see that it was not his name on his shirt but that of F. C. Phillips, who was attending the Exhibition as the nephew of Charles F. Brush, the American arc lamp pioneer. Phillips was in the next room, and the laundry had switched their clothing. In a sense Phillips was a rival, as Brush's British offshoot was promoting the incandescent lamps of St George Lane Fox. But there was no suggestion of foul play.

Joseph Swan made three visits to the Exhibition, writing home enthusiastically to his wife about the success of his lamps. 'Last night in the *Salle du Congrès*,' he wrote on 23 September, 'there was a meeting of the Society of the Telegraph Engineers. We, of course, lighted the hall with our lamps. There was only one opinion as to the manner in which it was done. We turned the lights in and out to accommodate the lecturer, who had magic-lantern demonstrations, and this was done quite as promptly as if gas, instead of electricity, had been the lighting agent.'[3] For the pioneers of incandescent lighting the ability to turn the lamps on and off easily and to dim them if necessary by reducing the current was a significant advance.

Each of the light bulb exhibitors had been allotted a space, and Swan had the buffet, of which he wrote on 24 September: 'Everything at the Exhibition has so far gone very favourably for us. Everyone is complimentary in the extreme. Certainly our rooms look well. The buffet is exquisite so far as our light is concerned.' The special feature of the buffet lighting was a kind of chandelier Edmunds had had made in London by the firm of Faraday & Sons. 'I remember telling Mr Harold Faraday,' Edmunds recalled, 'that I wanted fittings that could not be mistaken for gas, or oil, or candle or any of the other old fashioned methods of illumination... Hundreds of small spherical lamps were made at Newcastle with platinum loops, which were hung on to spring hooks, as a lampholder, to which were attached the insulated wires. To display these Mr Faraday designed a fairy-like structure of light lacquered brass... the whole structure was extremely decorative and effective for exhibition purposes.'

The greatest coup Edmunds had in Paris came through contact he made with a fixer called Maurice Simon, whom he described in his 'Reminiscences', a little patronizingly perhaps, as a 'splendid Frenchman'. Through Simon, Edmunds met Charles Garnier, architect of the brand new Opera House, who expressed an interest in the Swan lamps as a replacement for the existing gas lighting. Edmunds managed to get permission from the Exhibition organizers to take on this extra-mural project. Maxim's United States Electric Light Company got to hear about it and put in a request to take part. They got the job of replacing the gas footlights, while Edmunds had the more difficult, but prestigious, task of lighting the whole auditorium and the stage. He had to beg and borrow equipment, getting his generators for 600 lamps from Siemens and his cables from a French company.

To his surprise Edmunds found the engineers who looked after the gas in the Opera House very helpful. Perhaps they thought he was doomed to failure. Certainly Garnier insisted that the gas fittings stayed in place as an emergency measure. 'It was an interesting example of the new light struggling to assert itself in the face of the existing gas illumination,' Edmunds recalled. 'One could only work during certain limited hours, passing through mysterious trap-doors in the ceiling. The electric lamp bulbs had to be carefully attached and wired, so that they would be in their proper electrical circuit, as well as suitably supported under these trying conditions.'

All went well, and what was perhaps the first ever performance of Rossini's *William Tell* under electric light went by without a hitch. The demand for Swan lamps in France soared and the company began manufacturing there, once again recruiting German glassblowers to create the bulbs. Edmunds himself was called off to the port of Brest to demonstrate the bulbs to the French Navy. From there he went on to Barcelona to establish Swan patents in Spain.[4]

There was no doubt that Swan had a good Exhibition. He, personally, was awarded a Gold Medal and made a Légion d'Honneur. However, there was a cloud that hung over the otherwise brilliant display in the Palace de l'Industrie in the Champs Elysées. Edison had begun a patent war, which was occupying Swan's mind. His men had tried to get the Maxim bulbs banned from the Exhibition on the grounds that they infringed his patents in America. Not long afterwards they abandoned legal action and simply *bought* Maxim's patents. Swan knew he had a strong case for his own invention, but it did not help when the final honours were handed out that Edison achieved the highest accolade of all, the Diplôme d'Honneur. And when a jury judged the performance of the four lamps, Edison came first.[5]

It is little wonder, then, that the Edison camp crowed about their success and set about forming French companies to manufacture and market their technology. Their 'system' was also taken to Germany, where in time it formed the basis of one of the largest electrical companies in that country. And, inevitably, the Edison Jumbo generator was soon heading for London to challenge the British pioneers of electric light. This new industry was truly international. However, a long way away from the rivalry and excitement of Paris, history was being made in a charming English country town which, by happy chance, appeared to be ideally suited to pioneer the new system of lighting.

CHAPTER 4

THE OLD MILL ON THE STREAM

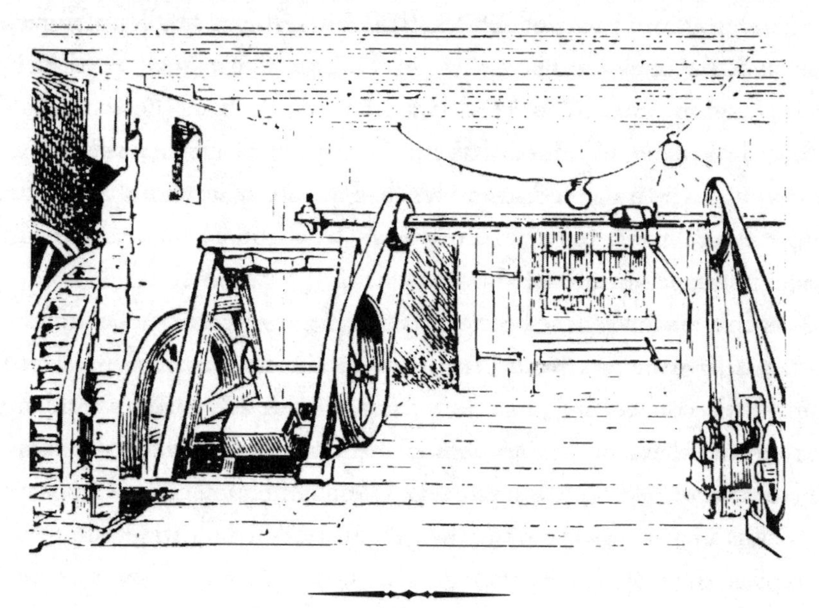

If you scrape away the layers of weathering and graffiti from a small brass plaque fixed to the wall inside the eccentric 'Pepper Pot' building in the main square of the pretty Surrey town of Godalming, it reveals a dedication which is so startling as to be barely believable.

THIS PLAQUE COMMEMORATES
THE WORLD'S FIRST PUBLIC ELECTRICITY SUPPLY
ON 28th SEPTEMBER 1881

NEAR THIS SPOT WAS INSTALLED
THE FIRST ELECTRIC LAMP TO LIGHT THE PUBLIC HIGHWAY
UNDER THIS PIONEERING DEVELOPMENT

The plaque was put up to commemorate the 100th anniversary of the

Godalming scheme, in recognition of the fact that this was a very early – if not the earliest – example of public electricity supply, not only in Britain but in the world. There were, in fact, other contenders, notably in the Derbyshire town of Chesterfield, as well as a small-scale operation in San Francisco. The belief that Edison's New York Pearl Street power station, opened in 1882, was the first such in the world (still repeated in many accounts of the early days of electricity supply) can be discounted. In fact Edison's men had a station working in London before Pearl Street, but this too was switched on a good year behind Godalming and Chesterfield.[1]

It is only since we have become accustomed to electricity being transmitted long distances from giant power stations that it seems bizarre that a little country town should claim to have achieved a world first in the early days of the technology. All the early electricity supply schemes were necessarily small scale. Godalming's claim to a place in history is that it was the first town to turn off its gas street lamps and to replace them with electric lights, at the same time offering a supply of power to any shops or businesses or householders who could afford it.

Like many towns in Britain that were supplied with gas for lighting by a private company, Godalming's council haggled every year about the cost of the annual contract. The company wanted to do the best it could for its shareholders, who were often prominent local people, and the council wanted the best price for the ratepayers. Though, when the industry was founded in the 1820s, there had been rival gas companies trying to compete in the same district, the absurdity of rivals laying two sets of pipes on one road soon put an end to the free market. In effect, gas companies had a local monopoly. For that reason quite a number of town councils had taken to running the gas works themselves. However, it was more common in the smaller towns, like Godalming, for the council to rely on a private supplier.

The councillors of Godalming took their historic decision to experiment with electric street lighting not because they regarded it as something wonderfully new or superior to gas but because it promised to be cheaper. The man who gave the town the chance to break its gas monopoly was the youthful proprietor of J. and R. Pullman, leather dressers, established in the old Westbrook Mill, which took its water power from a side channel of the River Wey. John Pullman had inherited the firm in 1873 when he was just twenty-four years old and had made a go of it. He was more of a City of London man than a Godalminger, with many interests in the old Square Mile, and he would have had the opportunity to see electric lights at work in many places by 1880.[2]

Westbrook Mill had two working waterwheels. Whether these were still in use for the leather-dressing business is not clear. If they were, then Pullman reckoned there must be spare capacity. They were extremely efficient waterwheels of a kind that had been devised in the 1820s by a French mathematician and inventor, Jean-Victor Poncelet. These wheels had curved blades that made them spin like a turbine and were said to be twice as powerful for a given flow of water, as more conventional 'undershot' wheels turned by water rushing below them.[3]

Pullman proposed that he would make his Poncelet wheels available at no charge to drive a generator. This would provide enough power to light the centre of Godalming and some side streets and to offer a supply to shops and households. In return Westbrook Mill and his private office and rooms would be supplied with electricity free of charge. We do not know for certain, but it seems very likely that it was Pullman who invited a London supplier of electrical equipment, Messrs Calder & Barrett, to bid for the franchise to light Godalming. They offered to light the streets of the town for £195 per annum, £5 less than the gas company had bid. The council wanted the gas company to continue to provide street lighting for one month while they experimented with electricity. It refused, saying it would only offer a supply for a minimum of three months.[4] The loss of street lighting was not, anyway, exactly devastating for Godalming Gas Light and Coke Company. Established in 1836, it still had plenty of customers, including the town's railway station, with which it had negotiated a contract to run for 100 years. Godalming station finally went electric when the contract expired in the 1960s.

As the Godalming lighting scheme was to be hydro-electric, it would be both cheap to run and free of pollution. A local gas works was a notoriously foul-smelling installation, and a dreadful place to work in the heat and grime of the coal-burning retorts. In contrast the Godalming electricity station, housed in the old Westbrook Mill, would hum along quietly and barely raise any dust. Siemens were chosen to install the generator along with their own arc lights and Swan's incandescent lamps.

As the days were drawing in that autumn in 1881 and the gas lights would not be coming on, there was no time to rig up any special lamp posts for electric lights. Westbrook Mill was on the far side of the River Wey from the town, and the mains cables had to be brought across somehow. No detailed history of the wiring of Godalming has survived, but it was certainly makeshift. Some details were provided in 1954 by eighty-six-year-old George F. Tanner, who gave to Godalming Library Services his reminiscences of the scheme. He had been born in 1868 and as a thirteen-year-old boy took a special interest in the new lighting as it was put up in September 1881. His father had a draper's shop in the High Street. Tanner recalled:

> My father was one of the first to have light in our shop and dining room and generally through the house ... The lamps were much as they are now but slipped into two brass slides like an inverted letter U. In those days we boys often had magnets to play with and the similarity intrigued me, so one day in our showroom when no one was about I took a needle to see if the electricity would act as a magnet and held it across the base of these two slides. The

> needle vanished and on my finger and thumb were deep white hollows where the needle had been. It had instantly fused. This was never done again as you can imagine.
>
> The wires were not insulated then. The dynamos were at Pullman's Mill and the river gave the power so the wires were brought overhead from there along the bottom of the Vicarage garden. At that time the wooden bridge was out of repair. The present brick bridge (which I remember being built) had taken its use and so it had decayed and become fenced in with a closed fence and the wires were carried along overhead of this, not very high up.
>
> There was opposition as you can guess to anything new and the story goes that two men with their cargo of beer came along one night and one lifted the other up to tear the wires down. But when he grasped them the current imprisoned both ...
>
> The story goes that old Mr Bridger who at one time was Mayor (or several times so) had shares in the Gas Company. He, it is said, liked his liquid nourishment. The arc standard by the Market House was loose and one night he was 'out to get one back' for the Gas Company and so embraced it and shook it and was heard muttering, 'B–b–b– 'lectric light!'[5]

The centre of town and the High Street were lit with arc lamps. These were not only offered at a lower rate than the existing gas lamps, they burned with much greater 'candlepower'. However, the Godalming scheme would probably not have been attempted if the new Swan lamps had not been available. It was these that John Pullman wanted in his office and dining room at Westbrook Mill as well as the interior of his works. For the public, though, it was the arc lights in the centre of town that drew the crowds. They were first switched on towards the end of September 1881. Among the Fleet Street newspapers that came to take a look was the *Graphic*, which illustrated its report published on

12 November with a dramatic sketch of the Pepper Pot and town square at night:

> The pretty little town of Godalming has gained for itself a distinguished place in the history of modern scientific developments by being the first town in England which decided upon the bold step of substituting the electric light in the place of gas for lighting its public streets... The effect of the quaint old High Street, with its gabled houses and miniature Town Hall, lit by electric light, is so strangely 'theatrical' that one almost expects to see a bevy of fair damsels appear from the 'sides' and dance across the street, while the 'heavy villain' of the piece is attempting to conceal himself in the deep shadow at the back of the Town Hall.

Meanwhile in Chesterfield, Derbyshire, a dispute between the council and its private gas supplier threatened to return the town to the Middle Ages. On 27 September 1881, *The Times* reported:

> For some weeks past the town of Chesterfield has been in darkness owing to the Corporation being unable to come to terms with the gas company for the continued lighting of the streets by gas. It appears that the latter have raised their price, an arrangement which the Corporation refuse to accede to, consequently the lamps have been removed, and the town is in total darkness every night, much to the danger and inconvenience of the inhabitants. Accidents at night are of frequent occurrence, and an indignation meeting has been held, but neither the Corporation nor the gas company show any signs of giving way.

As an emergency measure the Watch and Lighting Committee at Chesterfield ordered fifty petroleum or 'gas oil' lamps to be put up. They would each burn for eighteen to twenty hours a time and would cost a lot less than gas. However, a chorus of anxiety was recorded by the newspapers: women being molested in the murky streets of the town. Word had got round about the new electric lighting and pressure built up for the council to call in an electrical lighting firm. Just a few miles from Chesterfield an engineering works was already lit with arc lights by a company run by one of the most influential of electrical pioneers, Robert Hammond (about whom much more later). Hammond was promoting the arc lighting system of the American Charles Brush, and was only too happy to put in an installation wherever it was required. When Chesterfield made it known it wanted to experiment with electric street lights, Hammond put in the successful bid in competition with Calder & Barrett, who were hired for the Godalming scheme, and the French-owned Société Générale d'Électricité. The committee chose Hammond's scheme on the understanding that this was a trial rather than a permanent installation.

As a temporary measure to relieve the gloom in the centre of town, Hammond's company immediately put in eight Brush arc lamps, powered by a temporary generating station in the old Theatre Royal. While the larger scheme was being installed, Chesterfield had to make do with the gas oil lamps hung in the streets. Hammond had met the Watch Committee on 1 October, and the temporary lighting was working a week later. A more permanent installation would need a fixed central station and many more lights. Hammond's company had bought the patents of St George Lane Fox, who had exhibited incandescent lamps in Paris, and these were tested in Chesterfield by the inventor himself in late October. Hammond's scheme was very ambitious for the time: nine miles of streets to be lit with more than twenty arc and 100 incandescent lamps, supplied with power by dynamos feeding current through fifteen miles of wire. The generators were driven by steam engines that drew water for the boilers direct from the River Hipper, which runs through the centre of Chesterfield. There were teething troubles, but the new lighting was operating satisfactorily by the summer of 1882.

When the Godalming scheme was running, much was made of the fact that the lights were powered by a watermill which turned a generator. *Punch* inevitably had a bit of fun: an item headed 'FIRE WATER' read: 'Godalming has achieved a triumph. Its fireworks are waterworks, and the little town is electrically lighted by water power.

'"How do you do it?" inquired a simple stranger. The Godalminger took him down to the water's edge and fully answered the question by saying, as he pointed to the river,

'"That is the Wey."'[6]

It was, however, the *Daily Telegraph* that saw the future in Godalming. A leader writer, anticipating what he hoped would be an end to 'the despotic sway of gas', waxed lyrical about a future in which the world would be lit by the clever conversion of natural power into brilliant lighting. No more gas meters! No more monopoly gas companies with extortionate fees!

> It has been reserved for the little Godalming to turn its river, the slender and rippling Wey, into a piece of machinery, and set it, just like any other mechanical servant to the task of lighting the streets... The days when gas companies can pump into our houses a noxious, explosive vapour like carburetted hydrogen, through uncertain machines called meters and charge an abnormally extortionate price for it are numbered ...We shall not want the stoke and the collier so much if only the example set by the good people of Godalming be followed.
>
> The waterfalls, millheads and rivers will quietly be making all our electricity by day and we shall be consuming it as easily at night, or the winds and tides will be made to labour for us. Nature in all her varied moods will be called in to help us fight against the dark, and we shall be able eventually to turn night into day by the bright lights which Nature herself kindles for us.[7]

As it turned out it, it was another century before wind and water power were taken seriously in all but the most mountainous parts of Britain. And in Godalming the dream of lighting the town with the 'slender and rippling' Wey was soon dashed by a long period of torrential rain. The Poncelot wheels would turn only sluggishly when the river level rose and there was no longer a drop between the mill race and the outflow. It is not certain exactly when Siemens brought in a steam engine as back-up for the millwheels, but it was certainly needed quite early on. Small-scale hydro-electric schemes were not the future that the *Daily Telegraph* had imagined.

There were other problems with the Godalming scheme. It was found that the Swan incandescent lamps glowed brightly enough in Pullman's works and rooms, but were dim in town and were compared unfavourably with the gas lamps they had replaced. This problem of loss of current was soon solved, but it did not recommend the new lighting to private customers in the town. Calder & Barrett pulled out of the scheme, leaving it to Siemens. It had begun as a loss-making venture, but would break even and then become profitable once new subscribers offered to pay an annual fee to join the scheme. There were no meters to gauge how much electricity each household used, and the amount of power that could be produced was limited, but Siemens believed Godalming could break even and then turn a profit. It all depended on the size of the market for the new incandescent light bulbs.

The lighting of Chesterfield was more ambitious than that of Godalming. Its centrepiece was an arc light in the market-place atop a twenty-one-foot-high post, which provided an estimated 2,000 candlepower. In the spring of 1882 the *Electrical Review* reported that it was possible to read a newspaper in the light cast by this power lamp. However, when the carbons burned out they were not always replaced as quickly as the townspeople and the police would have liked. In fact the whole installation was makeshift and would clearly remain so until lighting companies, like Robert Hammond's, were given the go-ahead to lay their cables under the street like the gas companies.

But the electricity industry appeared to be on the point of take-off as more and more companies were formed. Quite a number had petitioned Parliament for individual Acts that would give them the right to lay their cables in a designated town or district. By the spring of 1882 there were so many new schemes for electric lighting, in the City of London and in provincial towns, that the government felt it would be best to introduce one 'catch-all' piece of legislation to govern the way in which the industry would be allowed to develop.

CHAPTER 5

A DIM VIEW OF ELECTRICITY

After his triumph at the 1881 Exhibition in Paris, and a further display of his electrical system in London's Crystal Palace early in the following year, it looked as if Thomas Edison's men were going to take Britain and Europe by storm. In March 1882 his representative in London, Edward Johnson, was able to create a new enterprise, the English Electric Light Company, with backing of £1 million and a board of directors that included eminent engineers and aristocrats. The politician Sir John Lubbock, who had been chairman of the London County Council, and William Preece, electrician to the British Post Office, were among the dignitaries who gave the enterprise gravitas.

Shares were issued, and an agreement made to pay £20,000 to the Edison company for British patent rights and equipment brought in from the Crystal Palace Exhibition. A site was found for this pioneer power station below Holborn Viaduct, which straddled the valley of the Fleet River close by Smithfield meat market. The viaduct had been constructed to ease the flow of horse-drawn traffic heading into Smithfield and the City of London, which had struggled with the steep inclines into the valley. It was a favourable location on the northern boundary of the City of London because a good deal of cabling could be run under the viaduct to the street lamps and hotels and restaurants along the road above without digging up the highway. This was important because no company had the right to excavate a public highway without special parliamentary permission. Both the gas companies and those laying down horse-drawn tramways had had to seek Acts of Parliament before they could break up the roads.

At the same time that the Holborn Viaduct station was being constructed, Edison was working to get the first central power station working in Manhattan. This Pearl Street station, which was to start operations shortly after Holborn, was the first of many established in New York by Edison and sparked the very rapid electrification of major American cities. The technology was substantially the same as that at Holborn, but the fate of the English station was very different. Whereas Americans embraced electricity with enthusiasm, English politicians took a rather cautious and dim view of it.[1]

Edison's men were confident enough at first, shipping over two of his 'Jumbo' steam-driven generators for the Holborn station which were powerful enough to light 1,000 incandescent lamps. These bulbs were strung out along the viaduct and Newgate Street and fitted in hotels and shops and restaurants. Four hundred were installed in the General Post Office, where William Preece presided. Edison's lamps also appeared in the City Temple, built in 1874 and made famous by the preacher Joseph Parker. The City Corporation was offered free street lighting for the first three months, with an assurance that the cost, thereafter, would not be more than that of gas lighting.

The Holborn Viaduct station generators first began to hum on 11 April 1882; the lights came on and all looked well for the future.

Careful records were kept of the costs of fuel and the level of demand for power. It was vital that the cost to customers was as close as possible to that of gas, for at this stage in the development of electricity price was more important than the quality of light. This price constraint made it difficult for the new companies to balance their books, but there was an optimism that, as the technology was refined, there would be money to be made from electric lighting.[2] There were signs in 1882 that investors were keen to put their money into the new industry, and shares were buoyant. But there was one obstacle in the way of expansion which had to be removed if electric lighting was to compete with gas: the companies would need permission to dig up public roads to lay their cables.

As a result of the number of petitions to Parliament for Acts which would grant this privilege, the Liberal government decided to lay down some general principles in a new law relating specifically to electric lighting. Though the intention was to put the industry on a firm footing, the Electric Lighting Act which was passed on 18 August 1882 had the unfortunate, and quite unintended, effect of pulling the plug on the nascent electricity supply industry in Britain. Whereas Edison's local power stations proliferated in New York and other parts of America, the Holborn station ran into financial trouble and was closed down in 1886. In retrospect the 1882 Act was a crazy piece of legislation that visited the sins of the past on the technology of the future. It was framed principally to prevent the new electricity companies exploiting local monopolies as the early gas companies had done.

The Victorians were not by any means wedded to the principles of the free market and free enterprise where a public utility was concerned. In fact the Liberals, elected in 1880 with Gladstone as Prime Minister, favoured public ownership of a range of public services including gas, water, sewage and tramways. And the most enthusiastic advocate of what has been dubbed 'gas and water socialism' was Joe Chamberlain, President of the Board of Trade, the department which would administer the 1882 Electric Lighting Act. Though a parliamentary committee took evidence when the new law was being debated in the Commons, the Lighting Act was regarded by all the newspapers as essentially the work of Chamberlain.

Joseph 'Joe' Chamberlain was a very successful and wealthy businessman who had first become involved in politics as mayor of the great manufacturing town of Birmingham. Chamberlain was by birth and background a Londoner, the son of a prosperous boot- and shoe-maker, also called Joseph. They were a Unitarian family in which entrepreneurial flair was tempered with a strong sense of social duty. Chamberlain had some education, winning prizes at University College School, London, in French and Latin and Mathematics and Science. But he did not go to university, joining the family firm instead.

Born in 1836, Chamberlain was fifteen at the time of the Great Exhibition in Hyde Park, where a relative, his mother's brother John Nettlefold, was impressed by an American machine for mass-producing iron screws for timber joinery. Taking a gamble, Nettlefold paid out a huge sum of money for the patent rights. He brought Joseph Chamberlain senior in as a partner and investor. A factory was established in Birmingham and young Chamberlain was sent there to work with Nettlefold and to look after his father's interests. He proved to be a brilliant salesman as well as a good accountant and made a fortune for himself and others.

Birmingham was an industrial conglomeration which had grown at a quite shocking pace in the nineteenth century and was poorly served by the essential public utilities, notably the water supply and gas lighting. When, in 1873, Chamberlain was elected as Liberal mayor of the city, he set about reforming its services in a radical fashion. There was already a precedent for major cities to run their own gasworks: Manchester council had taken over from private companies in 1843. And that was the solution Chamberlain brought to the water and gas supplies in Birmingham. In 1876 he entered Parliament as one of the Liberal MPs representing Birmingham, and was appointed President of the Board of Trade after the Liberals' election victory in 1880.

Chamberlain's view of how electricity should be handled was outlined in a Board of Trade Memorandum penned before the House of Commons had even considered the evidence collected by the Select Committee appointed to consider the Electricity Bill. The memorandum was unearthed in the Public Records Office by Brian Bowers, formerly senior curator at the Science Museum in London. Though it was not signed, the influence of Chamberlain is evident. Dated February 1882, it sets out some basic principles that the new law should reflect.[3]

First: 'Every person should be free to make and use electric power on his own premises without interference, so long as he does not injure or annoy others.' Second: 'Every person should be free to supply electric power to others without interference, so long as he can do it without interfering with the streets or with the property of others, or with existing electrical systems, or otherwise injuring other persons.'

There was a general principle, reflecting the 'gas and water' socialist view, that if a local authority wanted to set up an electric lighting system it should be allowed to do so. In that event there would be no point in a private company seeking to compete with it in that area. However, if a local council did not want to provide electric power itself it could license a private company to do so. The 'franchise' to provide power might last for a 'short time', after which the council would be entitled to take over. The memorandum anticipates that some councils might be obstructive, preferring to keep their gas works, declining the chance to provide electric light, and refusing to allow a private company into their area. In these cases the council could be overruled by Parliament if it was thought to be in the public interest. Finally it was recognized that the existing gas companies presented a potentially powerful obstacle to the introduction of electricity because they might campaign against it or, alternatively, try to provide it themselves to perpetuate their local monopoly.

When it was passed, the 1882 Electric Lighting Act was a measure that reflected, to a large extent, the contemporary belief that the industry was still experimental and that its commercial future was uncertain. The Board of Trade could give a company a licence to set up a lighting system, provided the local authority of the district did not object, and this would be valid for seven years. It might be renewed, or not. There was no way of telling. In the view of Brian Bowers, this is how Chamberlain and his civil servants imagined things would generally get moving, with councils giving the go-ahead and commercial companies setting up their equipment to demonstrate how it would work. Where such an order was made, the council would have the option, after twenty-one years, of buying the electric lighting concern's equipment with a view to running the installation itself.

Chamberlain's Electric Lighting Act was cautious, but it was not intended that it should hinder the development of electric lighting. There would be activity everywhere, with the issuing of licences and provisional orders. And that is what appeared to be happening in 1883, when sixty-nine provisional orders were ratified as well as ten licences. But the new lights were never switched on. Investment money dried up, partly because of a general economic recession, but more specifically because electric lighting companies no longer looked a good bet. The early schemes at Godalming and Chesterfield had revealed both economic and technical difficulties. A seven-year licence was hardly much inducement to a capitalist to risk their money. And even twenty-one years was not that attractive: any business that was thriving would get back only the basic cost of its equipment if a local authority decided to buy it out.[4]

In 1884 there were only four applications for provisional orders, while those from the year before were nearly all on hold. Siemens had wanted to put Godalming on a proper footing, with underground cables and a more permanent generator, but the economics did not add up. So, in 1884, Godalming went back to gas lighting. The same happened in Chesterfield: gas lighting returned to the streets. In fact the only scheme which kept going from the earliest days was one that had been started in Brighton, Sussex, by Robert Hammond, who had the franchise for the American Brush Corporation.

How much Chamberlain's Act was to blame for the dimming of the lights in the mid-1880s is still disputed by historians. Those affected by it, such as Robert Hammond and Rookes Crompton, were adamant that the twenty-one-year clause kept the investors away. In his *Reminiscences* Crompton wrote: 'This clause was considered so unreasonable by those whom we were then asking to risk their capital in our new adventure, and was moreover so parochial in its districting of systems of supply, that no really large scale attack on the problems of public lighting was possible ...'[5]

A large number of the provisional orders that had been granted by the Board of Trade had gone to companies which were offshoots of Robert Hammond's Anglo-American Brush Light Corporation. He had installed electric lighting not only in Chesterfield and Brighton but in two other south coast towns, Eastbourne and Hastings. Hammond was at pains to point out in letters to *The Times* that electricity was more expensive than gas chiefly because it had not yet been provided 'wholesale'. Therefore most lighting installations had been put in at a loss in order to keep the costs in line with gas. Hammond, at least, had a go at getting the Brush equipment working. A great many other companies peddling the Brush equipment got no further than obtaining the go-ahead from the Board of Trade. They had been formed to buy the patent rights to Brush equipment but pulled back from installing any lighting equipment.

Millions of pounds had been invested in these companies in the belief that they would be profitable, when in fact there was no way anyone at the time could hope to make money out of providing electric lighting. Small-scale installations could never be cheaper than gas, and the twenty-one-year 'buy-out' clause did not help. Nor did the details of the Board of Trade regulations when they issued a provisional order. The districts defined by the Board were not only very small, not much more than parishes, they were divided into two distinct locations, A and B. Any company awarded an order had so much time to bring in an experimental scheme in district A and a slightly longer time to serve the whole district by adding in B. The Board of Trade explained that they wanted to give companies time to set up a scheme but not so much time that a council might apply for an order and then do nothing. The public were to be protected, too, from exploitation by firms which insisted that only their own equipment could be used in their home. Householders had the right to choose which lamps and other fittings they wanted. This was a blow to companies that had spent a fortune on patent rights to a particular lamp.

In short, the Chamberlain Act set out to rein in a potentially powerful new industry which it feared would exploit the general public, only to discover that the beast it wanted to control was financially and technologically feeble. As a result the digging up of roads all over the country to lay electric cables alongside the gas pipes simply did not happen. Instead only those schemes in which electric wires could be strung out without interference with the roadway could go ahead.

Meanwhile investors in the many companies that had applied for provisional orders lost their money. Because so many were offshoots of the Anglo-American Brush Light Corporation – there was Metropolitan Brush, South East Brush, Midland Brush, Provincial Brush and so on – the collapse of the firms was known as the 'Brush Bubble'. It was compared with Tulipmania and the South Sea Bubble. (It was not very different from the more recent Dot-com Bubble.) The American who gave his name to the firm, Charles F. Brush, prospered, however.[6] He sold up in 1889 and abandoned the industry. At his palatial home in Cleveland, Ohio, he built a windmill-driven generator which provided him with all the electricity he needed. A company named after him still exists in Britain and makes turbo-generators.

The consequences of the Electric Lighting Act were unfortunate for the new industry in its early years and gave rise to many difficulties later on, as the principle of local provision of lighting ensured that electricity supply remained parochial when it would have been more efficient if supplied on a wider scale. But at the time the Act was passed, most of the leading lights of the industry did not envisage a future of big power stations supplying huge areas. Perhaps the most thoughtful contribution was made by William Siemens, who was of special interest to the Select Committee that considered the new law as it went through Parliament because he had been responsible for the lighting of Godalming.

Siemens was then fifty-nine years old, a member of the Royal Society, and a wealthy man. His was a calm voice in the clamour, and from the transcript of the hearings of the Select Committee you sense the members hanging on his every word. In 1874 he had bought a country house near Tunbridge Wells as an aristocratic retreat from his London home. The estate, known as Sherwood, extended for 160 acres and was an ideal place for the veteran of electrical installations to indulge in some domestic experiments.

Siemens described to the committee the comprehensive installation in his own private domain: 'I have a small 6-horse power steam engine driving two dynamo machines, which, during the time of the evening, give light to one arc lamp of 5,000 candle power placed outside the house to light the approaches and the terrace, and 30 incandescent lamps lighting the conservatory, drawing room, the inner hall and the dining room. When the electric light is no longer required in the house, it is used during the remainder of the night for horticultural purposes.'[7]

One of Siemens's great projects at Sherwood was the use of electric light to bring on, or 'force', plants, and he gave a number of talks on his night-time gardening experiments. In a collection of his published letters are two from a Tunbridge Wells neighbour, Lord Stratford, in acknowledgement of receipt of one of Siemens's home-grown melons. The second letter, dated 21 June 1880, reads: 'Lord Stratford's comps to Mr Siemens: He can now affirm on *personal experience* that the inside of the electric melon is quite as good as the outward appearance led him to expect…'[8]

That was not all. During the daytime, when no lighting was needed, electric power was used to pump water from the lower to the upper grounds and, on the estate's farm, to drive machinery for cutting up straw and root crops to feed to horses and cattle. Siemens also had an electric-driven saw for cutting timber. All these, he told the committee, effected a 'great saving in labour'.

At the time, as the title of the Bill before Parliament indicated, outside Siemens's own grounds, electricity was considered purely and simply as a new way of lighting streets, shops and the grander buildings. The idea that it might have a great many other applications was barely considered. When Siemens mentioned the use of electricity to power a tram, something the German branch of the firm had demonstrated in Berlin and Paris, the committee regarded the idea as irrelevant to its task. Yet one of the most intensive lines of questioning, not only of Siemens but of other prominent electrical engineers, was how much electric power cost. The many uses to which it was put at Sherwood, as Siemens pointed out, made it much more economical than if it had just been used for lighting. But the message appears to have been lost on the committee. It was a portent for Britain's later failure to make the most of power station resources.

In contrast to Siemens's electrified domain at Tunbridge Wells was the pioneer scheme of electric lighting in Godalming. The committee was naturally anxious to learn how it was going and what kind of model Siemens thought it had for the future. By the time he appeared before the committee to discuss this in late May 1882, the firm of Calder & Barrett had already withdrawn. The watermill had proved to be a fickle source of power, and Siemens, taking over the installation, had had to put in a steam-driven generator. He chose to place this in the centre of town, as the loss of current from the riverside power station was causing problems with the Swan lamps, which glowed a dull red and looked feeble alongside surviving gas lamps.

In answer to a variety of questions, Siemens said he thought that the major technological problems with electric lighting had been solved, but there remained a question mark over the economics of the business. Clearly Godalming was a loss-making venture, principally because no profit could be made just from the contract with the town council for street lighting: private customers would have to provide a substantial part of the income. He thought only 'eight or ten' private customers had so far taken the incandescent lamps, just fifty-seven lights in all. He was asked how many customers he would need to make Godalming, or an electric lighting scheme anywhere else, a commercial success.[9]

The answer Siemens gave reflected what most of the leading electrical pioneers believed at the time: an electric lighting enterprise would ideally cover an area of about 'a quarter of a mile square'. Although the number of houses would obviously vary from one place to another, his estimate would be 1,500 served by one power station. These would be the larger houses, for Siemens regarded electric light as something of a luxury. However, if electricity were supplied to smaller houses in a city the size of Manchester, for instance, there might be forty or fifty different power stations for the whole metropolis.

This view, that each district and small town would have its own independent electrical supplier, was shared by Thomas Edison, Rookes Crompton and most of the pioneers. It was inevitably influential in the framing of the Bill as it went through Parliament. Siemens made it clear that he thought the competition with gas lighting was overplayed. When he was asked to what extent gas lighting might be replaced by electric lamps, he replied: 'That I could not say, because at present in our drawing-rooms and dining-rooms we have practically banished gas already, therefore the electric light will come in mostly in lieu of candle lighting and oil lamps.' The future for gas, he thought, was in heating homes and factories, for it was much cheaper than electricity.

When investment in lighting schemes collapsed, Chamberlain was inevitably blamed by the newspapers for framing legislation which was overly biased in favour of public control. This was not entirely unfair, as the Electric Lighting Act was deliberately antagonistic towards private companies. But his vision of what the future might be like was not very different from that of William Siemens. They were both wrong about the future of electricity, which was soon to develop in a way neither had anticipated.

CHAPTER 6

A PRE-RAPHAELITE POWER STATION

There was a loophole in Chamberlain's electric lighting law: if you could create a supply system without digging up the highway you were free to transmit power wherever you liked. This allowed one enterprise to flourish when all about was gloom and stagnation in the young industry. It was the most exquisitely British creation, involving a fashionable West End art gallery, a young man descended from the Doges of twelfth-century Venice, and the Earl of Crawford and Balcarres. The fashionable shops in New Bond Street also played their part. This is how it came about.

The debonair and very handsome Sir Coutts Lindsay, heir to a small fortune, decorated soldier and veteran of the Crimean War, left the army in 1862 to devote himself to painting. His first wife, Caroline Fitzroy, whose mother was a Rothschild, was also a painter and a poet. Both were said to have considerable talent and they were sympathetic to a group of painters formed in the late 1860s who called themselves the Pre-Raphaelite Brotherhood. They disliked the direction painting had taken since the influence of Raphael and others in the sixteenth century. The Brotherhood included Holman Hunt, Burne-Jones and Dante Gabriel Rossetti.

When the Royal Academy made it clear that it was reluctant to exhibit the works of the Pre-Raphaelite painters that Sir Coutts and his wife admired, they decided to create their own fine picture gallery in London's West End where they could be shown. With the backing of the landowner, the Duke of Westminster, they got permission to demolish a number of buildings in New Bond Street to create a site for an entirely new gallery.[1]

The opening attracted 7,000 visitors who paid an entrance fee of one shilling, making the gallery an instant success. It continued to win praise for its exhibitions into the 1880s. By that time there had been many experiments with electric lighting and it had proved especially suited to libraries and galleries where the pollution from gas lighting could be harmful. A cousin of Sir Coutts was the Earl of Crawford and Balcarres, the scientist and astronomer who had been the British Commissioner to the Paris Electrical Exhibition of 1881. He persuaded Sir Coutts to give electric lighting in the Grosvenor Gallery a try. An experiment with a small generator and some arc lights late in 1883 was successful and attracted many visitors who were doubtless as interested in the lights as in the paintings.[2]

Local shopkeepers were impressed and wondered if they might just take a feed off the gallery circuit for themselves. Sir Coutts obliged and the demand grew rapidly as more and more shop owners asked for a connection. The wires linking his new-found customers with the generator were strung out over the rooftops. For permission to extend his supply system all that was needed was a 'wayleave' from the owner of any property over which he needed to suspend his wires. It was perfectly legal and bypassed the irksome 1882 Act. However there came a point where the Grosvenor Gallery supply had spread its rooftop web of wires so far that it began to run into serious technical problems. Telephone subscribers began to complain of interference and lighting customers were disgruntled with the increasingly frequent breakdowns.[3]

The whole thing had to be overhauled, with the installation of more powerful generators needed to produce higher voltages which could take current to the more distant customers. To accommodate these steam-driven Siemens generators a large area had to be excavated below the Grosvenor Gallery. These improvements brought with them new problems. Letters began to appear in *The Times* complaining about the vibrations from the steam engines and there was further interference with telephone lines. By 1885 it looked as though the Grosvenor Galleries' bold experiment as a local power station was doomed to failure.

There was just one glimmer of hope. A young man with an Italian name had supplied the gallery with his own design of electric meter, and he had, from time to time, discussed with Sir Coutts the problems the gallery power station was having. The Earl of Crawford had heard that this young man had gained a reputation as a prodigy in the field of electrical engineering. He invited him to dinner to discuss the possibility of him taking over the defunct installation with a view to getting it going again.

For Sebastian Ferranti, who was then just twenty-one years old, the opportunity to take over the running of a power station was a dream come true. From boyhood he had formed the ambition to become an engineer, and had only ever interested himself in mechanical and electrical things. He was born into a family with a pedigree every bit as exotic as the name with which he was christened: Sebastian Pietro Innocenzo Adhemar Ziani de Ferranti. His mother, Juliana, was the daughter of a well-known portrait painter, William Scott, who had studied with Millais and had moved from London to establish a studio in Liverpool. His father, Cesar, was the son of a famous guitarist who had settled in Belgium. It was there that Cesar became fascinated by photography and established a business taking portraits of the wealthy, who were captivated by the novelty of instant images. The Ziani de Ferrantis could trace their ancestry back to one of the Doges, merchant princes, of Venice.[4]

Before she met Sebastian's father, Juliana already had four children by a previous marriage. She was an accomplished concert pianist when, at the age of twenty, she fell in love with a Polish guitarist, Stanislas Szczepanowski, who was living in exile in Europe. Brushing aside her father's warning that this man, twelve years older than her and already a widower, was not an ideal partner, Juliana married him in 1845. The couple then embarked on a concert tour of Europe, with Juliana giving birth to four children, each born in a different city. They were in Ostend in Belgium when the last child, Vincent, was born in 1852. It was the year Stanislas disappeared, abandoning them all to their fate.

Juliana returned to Liverpool with her children but continued to perform from time to time. In 1857 she was giving a recital in Ostend when she met Cesar, the photographer. They fell in love and he asked her to marry him. She could not divorce Stanislas because he had disappeared without trace, but the law allowed the marriage to be dissolved when he had been absent for seven years, which was in 1859. Juliana and Cesar nevertheless became lovers and she gave birth to a daughter in 1858, who was sent straight to a convent so that Juliana's father would know nothing of her transgression. The following year Juliana was received into the Catholic Church and she married Cesar in 1860. They settled in Liverpool, where Cesar set up a studio with William Scott and established a flourishing photographic business.

Sebastian was born in 1864, when his youngest stepbrother, Vincent, was twelve and the elder, Wladziu, sixteen. His stepsister Emilka was eighteen and the younger, Vanda, fourteen. When he was just two years old, both brothers joined the navy and Sebastian spent a good deal of time alone or with his mother or a nanny. His father had bought a property away from the centre of Liverpool, at Egremont on the far side of the Mersey, a river busy with both steam and sailing ships. From an early age Sebastian displayed an artistic talent, sketching ships and steam engines with precision and originality. His interest in mechanical things was encouraged by his stepbrother Vincent, who wrote to him from his ship, SS *Culmore*, and sent him stories about his adventures at sea.

Professor John Wilson, who has chronicled the history of the Ferranti business in great detail, suggests that it was Vincent Szczepanowski who set young Sebastian on the road to a career in electrical engineering. To pass his examinations to become a First Officer in the navy, Vincent had acquired a library of books on the working of steam engines and other machines, which Sebastian inherited. Vincent also took his young stepbrother to Spain on the SS *Culmore* in 1877. Later the two would work together, with Sebastian as the senior partner.

Sebastian's childhood letters, mostly to his mother, reveal that he was no scholar. What stands out is his gift for illustration and his determination when he was still a schoolboy to become an engineer. He was sent away to school in London when he was only ten years old. This was perhaps because home life was difficult, as Juliana's second marriage was not a success. Her first husband Stanislas had reappeared, much to the chagrin of Cesar, who became estranged from Juliana. It is perhaps just as well Sebastian was absorbed with his interest in mechanical things. He wrote to his father from his Hampstead school asking for 'a littel model of a steme fire ingun wakein by stime'.

When he was thirteen he was sent to St Augustine's, a Benedictine school in Ramsgate, Kent. Here, by good fortune, the headmaster recognized his talents and encouraged his experiments, providing him with a space to work in the school kitchens. With the science teacher, Father Burgh, he built a crude arc lamp powered by a battery. When Father

Burgh recognized his own limitations he introduced Sebastian to a Mr Jarman, a photographer and amateur electrician. During the school holidays Ferranti stayed mostly in London with his older stepbrother, Wladziu, which gave him time to visit museums and attend lectures. When he was sixteen he visited the newly formed Anglo-American Brush Electric Light Corporation, which fired his enthusiasm further. By that time he had mapped out a career for himself.

Though Cesar was not convinced Sebastian would succeed as an engineer, in 1880 he agreed to pay the fees for his son to attend University College London. It was then about the only academic institution which had an engineering department. But Sebastian did not stay there long. In 1881 his father suffered a stroke and did not fully recover. Cesar was separated from Juliana and was unable to keep his photographic business going. Sebastian, at the age of seventeen, had to make his own way in the world. In April 1882 he wrote to his stricken father: 'I know that I have the strength, the ability and the determination to overcome all difficulties.'

It was no great loss for Sebastian to leave university. In fact it enabled him to get his hands on electrical equipment quicker than he might have done otherwise. The Mr Jarman he had been introduced to by his science teacher had set up a laboratory in London and gave Sebastian space there. Here he designed a dynamo which he called the 'rabbit' and sold it for £5. However, he needed a job. The obvious place was Siemens, with their large works at Charlton in south London. He was given a letter of recommendation by the chief electrician of the South Kensington Museum (forerunner of the Science Museum). At first he was turned down, but he found a way to get an introduction to Alexander Siemens, the adopted son of William Siemens, who was in charge of electric lighting development made by the company.

Ferranti was still seventeen when he began to work for Siemens on improving their generators. His boss was Dr E. Obach, a leading technician at the time whose workshop had the most up-to-date facilities.

This was the perfect job for Ferranti and he excelled at it. Early on, Siemens trusted him to install their lighting equipment. It was while he was working on a scheme in Wolverhampton that he met by chance a friend of his father, Alfred Thompson, who was interested in the electrical business. Thompson recognized Ferranti's talent and introduced him to a successful company lawyer, Francis Ince. Among Ince's clients was Robert Hammond, whose company was lighting Chesterfield. Electrical engineering was a small world in the 1880s.

Ince and Thompson persuaded Ferranti that he would be much better off financially if he set up a company with them rather than working for Siemens. They were especially impressed by a new kind of generator that Ferranti had designed, and thought there would be a market for it. Ferranti decided to take their advice, but his sudden departure from Siemens was regarded by that company with considerable suspicion. No sooner had their young employee handed in his notice than he was patenting his lightweight yet powerful generator, which undercut Siemens on price. No action was taken against Ferranti, though William Siemens did write to *The Times* suggesting there had been a degree of foul play. But that was not the end of the matter. When Ferranti lodged his patent he learned that what he had invented was almost identical to a machine devised by the eminent scientist Sir William Thompson. The two did not fall out, but instead became firm friends. Thompson agreed to give up his rights to the generator in return for royalties of £500 a year, which was then a considerable sum of money.

The size of the royalty reflected the belief that this new Ferranti–Thompson generator represented a real breakthrough in the industry. It could give five times the power of any machine in existence and bring down the price of electric lighting so that profits could be made from it. Robert Hammond acquired the rights in the new invention, telling his shareholders in Anglo-American Brush that the new generator would allow electric incandescent lighting on a much greater scale than anybody had ever imagined. This was not a good time, however, to sell generators, though they worked efficiently. The 'Brush Bubble' was about to burst, as the British electrical industry was effectively switched off by the 1882 Lighting Act.

However, there was still some demand for electrical equipment, especially in Europe. With his generators unsold, Ferranti turned his hand to manufacturing a patented electricity meter, something the industry needed urgently, but had not yet perfected. He won a contract worth £5,000 to supply a company in Belgium and took orders from a German-owned Edison offshoot in Berlin. He also sold meters to the Crompton company, which was installing lighting in the Opera House in Vienna. And it was the meter business that gave him an introduction to Sir Coutts Lindsay's Grosvenor Gallery power station.

Ferranti was wined and dined by Lindsay and the Earl of Crawford and persuaded to take over as engineer-in-chief of the gallery on a salary of £500 a year. He took up his new post in January 1886, with £20,000 from the gallery's wealthy backers to completely re-design the power station to his own specifications. It was an opportunity unique in Britain at the time, and hardly matched even by the industry in America and Germany. It was not easy to find trained electricians in these early days, but Ferranti was able to recruit a young man called C. P. Sparks, who had studied at an electrical college founded by the ubiquitous Robert Hammond. Then he brought in his stepbrother Vincent to supervise the installation of new generators. These were 'alternators' of Ferranti's own design, and the largest and most powerful in the world.

In search of more recruits with the necessary expertise, Ferranti brought in H. W. Kolle, chief engineer of the Brighton station founded by Robert Hammond. There were others too with expertise in cable-laying and draughtsmanship. As the chronicler of Ferranti's history, Professor Wilson, has pointed out, this was a very young team. Vincent was by far the oldest at thirty-six, and none of the others had reached thirty. Sparks was the youngest at twenty-one, and Kolle was the same age as Ferranti at twenty-two.

There was one aspect of this new power station as it took shape in 1887 which made it radically different from most others that had been established up to that time. It was to run on AC (alternating current) rather than DC (direct current). Though the original Grosvenor Gallery system had also been AC, this was not favoured by Crompton or Edison. The great disadvantage of AC was that it could not be stored. With DC it was possible to charge up batteries, and the efficiency of these had been greatly improved by French inventors.

By happy chance, the Grosvenor Gallery system had been operated from the beginning with 'alternating current' or AC. This meant that the cables that carried a high current over the rooftops need not be very heavy, an important consideration. However, the AC system had to be at work continuously if there was to be any power at all. In every country where electric lighting was being developed there was a division of opinion between the AC and DC camps, known generally as the 'battle of the systems'. In America, it was fought bitterly with Edison claiming that AC was inherently more dangerous than DC. In Britain the disagreement was much less fractious, so that a DC man like Crompton would happily have dinner with the AC advocate Ferranti, and they were in fact part of an informal group of electrical engineers who called themselves the 'Dynamicables'.[5]

Ferranti and his team were committed AC men, as they had to be if the plans being laid for the Grosvenor Gallery station were to be successful, for the idea was to greatly extend its catchment area. With the technology then available, no DC system could be operated on the scale Ferranti envisaged, and the level of voltage or 'pressure' he required to transmit current for more than a mile was regarded as quite frightening. However, that had to be the way forward for electric lighting systems, because it was only when one power station could serve hundreds of paying customers that it could make a profit, and the price of electricity would have to fall significantly before large numbers of people could be induced to buy it.

This was recognized by all those promoting electricity at the time, among them Robert Hammond, whose experience went back to the earliest schemes in Brighton and Chesterfield. Despite the many setbacks he had experienced, Hammond continued to promote electricity enterprises around the country, arguing all the time that the only reason gas lighting had not been superseded was that it was produced 'wholesale' whereas electric power was still produced on a small scale. In 1884 he published a little book, *The Electric Light in Our Homes*, in which he told the following story:

> At the close of one of my lectures in London, a director of a gas company said he would like to ask me a test question: 'Are you prepared to light the five street lamps in your grove with electricity, from the dynamo at your house, at the same price as the gas company is at present charging for gas?' To which I replied: – 'Yes; I am prepared to supply these five lights at the same price as that charged by the Gas Light and Coke Company, provided that the gas company allows me also to do the remaining 250,000 lights, by the supply of which they are enabled to keep the cost of five lights so low.[6]

This 'wholesale' production of electricity was Ferranti's ambition as he brought the Grosvenor Gallery power station back into production. Before the station closed there had been just 300 lamps supplied. Under the new name of the London Electric Supply Corporation (LESCo for short) there were 11,000 lamps supplied by October 1887, just two months after the new generators were working. The transmission wires supported on special brackets designed by Ferranti spread across the rooftops of the West End with amazing speed, bringing electric light to clubs and theatres and opulent homes. The cables reached to the east of New Bond Street as far as Lincoln's Inn Fields, nearly a mile away, and en route lit up some buildings by Trafalgar Square. As the roads were never disturbed, no local authority could intervene and the Board of Trade had no powers to stop the expansion of the remarkable Grosvenor Gallery grid. A real coup was the wiring up of Marlborough House, the London home of the Prince of Wales, adjoining St James's Palace on Pall Mall.

While dozens of would-be electricity supply companies still languished in 1888, LESCo attracted an astonishing £1 million in private investments. Among the twenty-eight shareholders were Sir Coutts and Lord Crawford, but the biggest risk-taker was a younger brother of Sir Coutts who had been elevated to the peerage in 1885 as Lord Wantage. Another Crimean veteran, Wantage had won the Victoria Cross for bravery at the battles of Alma and Inkerman, had been a Conservative MP and had married into a fabulously rich family. He risked £220,000 in Ferranti's LESCo venture. In 1889 James Forbes, a very experienced manager and businessman, was brought in to provide some stability for the company. Forbes, then in his early sixties, had begun his career working on the Great Western Railway with Isambard Kingdom Brunel and had been involved in the management of many different companies.

When the plans Ferranti had for LESCo were first reported in the technical press they were regarded as incredible. He had persuaded his backers that he could build a power station miles away from all their customers in central London. This is what the Gas Light and Coke Company had done, setting up their works on the River Thames to the east, on a site that was named Beckton after the managing director John Beck. Coal was brought in from the north-east of England in collier ships that could deliver direct to the works. From there gas was piped into London under pressure. Inspired by this example, Ferranti found a site at Deptford, on the south bank of the Thames, on which to build what would be at that time far and away the largest electricity power station in the world. The Grosvenor Gallery installation would become a sub-station where voltage was transformed for distribution to customers.

Ferranti's Deptford power station, with its huge steam engines and twenty-ton generators, was designed to send a massive 10,000-volt current the seven miles into London. He designed his own high-tension cables because there was nothing available that could handle such pressure. And he had to convince the Board of Trade that if one of these cables was severed it would not cause a disastrous accident. Fear of electricity was still commonplace and was not unjustified: there had already been a few fatalities in the industry, even with much lower voltages. A dramatic demonstration of the safety of the 10k cable was staged at Deptford in 1890, as the works were nearing completion. As anxious Board of Trade inspectors looked on, Ferranti's assistant, Kolle, held an uninsulated chisel on the live cable while the site foreman, H. W. Henty, brought a sledge-hammer down on the chisel head. To everyone's relief, the main fuse blew as soon as the chisel hit the live part of the cable and Kolle was able to walk away unharmed. Asked if he had been frightened, Kolle is said to have quipped: 'I was scared out of my life. Young Henty had never used a sledgehammer before.'

The old New Bond Street power station was still operating, with its wires strung out as far to the north as Regent's Park and to the east to Lincoln's Inn, when Deptford was ready to go into action. With the help of James Forbes and his former railway connections, LESCo had negotiated to bring the 10k cables into town along the private property of the railway lines, by passing the provisions of the 1882 Act.

However, LESCo could not avoid the law altogether. The days when electricity cables could be carried across the rooftops were clearly numbered. Transformers would step down the high voltage coming into London and from them the power would be offered to consumers over a very wide area. Roads would have to be dug up, and that meant applying to the Board of Trade for a provisional order, which in turn would involve the local authority of any district. To complete Ferranti's ambitious scheme, LESCo put in for a total of twenty-six such orders. The company did this just as the 1882 Act was being amended to answer bitter criticism made by Rookes Crompton and others of the 'buy-out' clauses which had inhibited investment in the industry. After 1888 the tenure for a company before a council's right to buy was doubled, from twenty-one to forty-two years, and the price to be paid for an enterprise had to be 'fair' rather than 'scrap value'. The effect was to unleash a pent-up demand for provisional orders.

The scramble to provide London with electricity, and the unprecedented scale of Ferranti's plans, prompted an inquiry by the Board of Trade. From their Inspectorate they appointed the ideal man, a well-known football referee, Francis Marindin, who could be relied upon to take a dispassionate view of the competing interests. Eton-educated and a former officer in the Royal Engineers, Marindin's expertise was in railway safety and the rules of association football.[7] Electrical engineering does not appear to have been one of his subjects, but the issues on which he had to report and make a judgement were only partly technical. The most important considerations for Marindin were to keep to a minimum the disturbance to the public caused by the laying of cables and to prevent one company getting hold of a monopoly of supply over a large area.

This was not good news for Ferranti, as the whole point of the Deptford project was to bring down the price of electricity by supplying it to a large number of customers from a single central source.[8] Counsel for LESCo argued the best possible case. The power station at Deptford would get its coal supplies delivered on the Thames, so there would be no need for wagons to carry them to boilers sited in the busy West End. Traffic congestion would be eased. Marindin listened respectfully, but it was no surprise when he refused to allow LESCo the catchment area they wanted. It did not help that when he visited Deptford during the inquiry a boiler blew up, killing one of Ferranti's men and injuring two others. The thought that the young genius had become too ambitious might have crossed his mind. But it was really the determination to ensure competition between companies that persuaded Marindin to thwart LESCo's plans. Not only did LESCo get a much smaller supply area than it needed, Marindin ensured that it lost some of its West End customers to rival companies.[9]

Ferranti's companionable rival, Rookes Crompton, recognized this and wrote in his *Reminiscences*:

> The district that was allotted to the proprietors of the Grosvenor Gallery Company was unfair on them, as they were not allowed to operate over the district which they had been supplying previously by overhead wires. In place of this, by the Marindin Enquiry, they were allotted a very restricted area on the south side of the river, through which their mains would have to be laid from Deptford to the West End... In the West End they were put into competition with local central stations run on the D.C. system; and, as during these early stages, this system was much better understood and more easily worked, the competition very nearly put the supply by alternating current out of existence.[10]

There was, in effect, a kind of social class division in the supply of electricity: AC was for large numbers of not so well-off customers, DC for compact wealthy districts. As Crompton was ready to admit, Ferranti was simply ahead of his time.

Marindin's judgement, delivered in May 1889, was disappointing but not fatal for the company. Deptford power station as it took shape continued to be a technological wonder and was visited in September by Thomas Edison, whose reputation as a wizard had not diminished. He was lionized by the British press, who were anxious to know what he thought of Ferranti's monster power station. The correspondent of the *Daily News* asked Edison how he thought, in electrical terms, we were 'getting on in this old country?' 'You may be slow to begin,' replied Edison, 'but' – and here he nodded once or twice – 'I must say that when you do go ahead you may even beat us.'[11]

These were regarded as generous words from the 'wizard', though the judgement was questionable: Britain was not slow to begin so much as sluggish in its development. However, if Deptford worked, it would be true that London at least had got ahead. If it did so, as the *Daily News* pointed out, it would be in defiance of Edison's passionately held view on how electricity should be generated and delivered.

> It is useless to disguise the fact that Mr Edison's views on electric lighting are not in all respects in accord with those which have been practically adopted by the London Supply Corporation (LESCo) – in other words, by their accomplished engineer Mr Ferranti. Mr Edison's own system of lighting in America is the low-pressure system; the London Company's is the high-pressure; Mr Edison's is the system of direct current; the London Company's is the alternating current. 'Our New York low-tension wires,' said Mr Edison, 'are so absolutely safe that even a child may play with them.'

The safety of low-voltage DC supply was one of Edison's favourite topics. A few weeks before he left America for Europe he had been a star witness in a courtroom battle over whether or not electrocution would be a humane form of execution. A convicted murderer, William Kemmler, who had hacked his common-law wife to death with an axe, awaited the decision of New York State after the proposal to use an electric chair instead of the hangman's rope to kill him had been challenged. Edison's great rival in New York was George Westinghouse, who, like Ferranti, was an AC advocate. When he learned that it was proposed to use AC current to kill Kemmler, Westinghouse hired a lawyer to challenge the state's right under the Constitution to use electricity for an execution. His motive was to allay bad publicity about AC, which Edison assured everyone was much more dangerous than DC. Edison, of course, told the court that it would be fine to use AC to kill murderers: it would be quick and painless. The state won, and in August 1890 Kemmler was the first man in history to go to the electric chair. 1,000 volts AC failed to kill him. In the end, with a 2,000-volt AC current, he took eight minutes to die.

What Edison said about AC was largely nonsense, as Ferranti and Ince pointed out in a little monograph they published in 1889 entitled *The Dangers of Electricity*. This showed that there were more deaths with DC in New York than with AC in London. In fact, at the time, there had been no electrical fatalities at the Grosvenor Gallery station, which was operating at 2,400 volts, and the death at Deptford was unrelated to electric current.

When the *Daily News* reporter inquired further after the Deptford visit, Edison conceded it would 'go' but that Ferranti was 'putting all his eggs in one basket'. To demonstrate this 'he leant forward, and placing the nail of his thumb against the nail of his forefinger, he remarked that a flaw of a tiny fraction of an inch in the high tension main would in the twinkling of an eye plunge London into darkness.' This was the other DC defence: you could back up a power breakdown in the generators with batteries with direct current, but AC had to keep working continuously.

Edison was asked which capital city was best lighted at that time: Paris, Berlin, New York or London. He considered Berlin to be the best lighted town (electrically) on the continent. It had a station as powerful as Deptford, but half the size. New York had 700,000 lights, all provided by small local stations. It was agreed that London might be on the point of catching up as far as lighting was concerned. But in one respect it was a long way behind. 'The next thing you will have to do,' said Edison, 'will be to turn your electric force into motive power.' The *Daily News* reporter noted: 'Here, perhaps, Mr Edison was more in his element, than when discussing light. With a smile he said: "America is the country for that; in New York and other places, even small shopkeepers use electricity in scores of ways, in place of hand labour and old-fashioned mechanical appliances; they can buy it cheap – for a few cents for that matter."'

Far more significant, however, was the use of electricity for transport. 'Take electric railroads,' said Mr Edison; 'every small town in the United States is supplied with them – the small towns even more liberally than the great cities; we have thousands of miles of electric railroad, and the motive force supplied by overhead wires; and we find it cheaper and better in every way than traction by horses – whose working lives last only four years.' London did have plans for an underground electric railway, Edison was told, but he was unimpressed. The *Daily News* reported: 'What did surprise Mr Edison, was the obvious fact that the Metropolitan and District Railways were not driven by electricity: "Nothing would be simpler," said Mr Edison. "than to substitute electricity for steam." He had offered to do it long ago. If he got the order now he could carry it out almost offhand. And he gave a glowing picture of what the Underground would be without its steam and its vile choking sulphur fumes.'

Edison put London, and, by implication, Ferranti in their place. The capital might be about to get hundreds of thousands of electric lights supplied with current from Deptford's jumbo power station but, in its use of electricity generally, it was rapidly falling behind New York and other American cities as well as Berlin. To have trams still drawn by horses and steam-driven railways in a great city was, in Edison's view, a sign of backwardness. And where the metropolis might be just ahead of the game, with Ferranti's huge power cables and giant generators, what was being attempted was very risky. The *Daily News* reporter ended his piece with this anecdote: 'A gentleman of the party was remarking that if "once" a breakdown happened, public confidence in the new light would vanish. "Twice," said Mr Edison, on the instant, with his quiet smile; "the first time the public will excuse you, they will make allowance for inexperience; but if you do it a second time you are done for."'

Edison was more prescient than even he could have imagined. Though Ferranti had problems with his high-tension cables, which tended to be set on fire by the sparks of passing steam trains, the Deptford station was ready to send power into London by November 1890. The original Grosvenor Gallery power station was being converted into a substation, where current from Deptford was reduced by transformers for further distribution. It was not functioning because neighbours who complained about the noise and vibration the steam-powered generating plant made had an injunction to stop it. The hand-over to the new system had to be done very quickly, and transformers were set up on temporary woodwork frames. In fact the whole room in which they were housed was wooden and 'an accident waiting to happen'.

At 6.30 a.m. on Saturday 15 November, a workman operating a switch allowed a huge spark to flare and was so startled that he failed to cut the current off as he should have done. This started a fire that burned out the room along with the transformers. It was all over in twenty minutes. Nearly 40,000 lamps would not be lit that night. Two weeks later they came on but the repairs had been done too hastily and the transformers burned out again. The directors of LESCo decided to shut Deptford down until the problems could be sorted out. Work continued on upgrading the whole system, taking down the rooftop wires of the old Grosvenor Gallery network and putting them underground, the main cables finally being put in place and the largest generators ready for installation at Deptford. But the directors of LESCo felt they could no longer fund Ferranti's most ambitious plans and that Deptford would be operated at a lower level than he had planned.

Though LESCo continued in business and won back many of the customers it had lost, Ferranti's disagreement with his backers led to his resignation in 1891. They continued to operate as one of the largest suppliers of electric light in the capital, while he went back to being a successful manufacturer of electrical equipment. His factory making meters was still going and doing well. It was by no means the last that was heard of Ferranti – he was only twenty-six when he resigned – but it was the end for a brilliantly conceived scheme that was the model for all future electricity supply systems. It was a long while before that happened in London, or, indeed, in the rest of the country. With one notable exception, electricity supply was local, provided by a council or a small private company, and remained so for nearly half a century.

CHAPTER 7

A TALE OF TWO CITIES

'It must be with a troubled mind that one approaches the history of electric lighting in Leeds,' wrote a correspondent in the *Electrical Review* of 19 May 1899. 'It is a record which, for sheer vacillation and sheer incompetency, is the worst in the whole of the written pages of electricity.' This damning judgement on the way the proud Yorkshire town came to terms with the wonderful new technology of electric lighting is not entirely unfair. But the very same criticism could be levelled at a great many towns in Britain which puzzled for several years over how and why and when they might accommodate an electric power station in the Victorian fabric of the town. The Leeds story, farcical though it is in many ways, is representative of a particularly British approach to what newspapers of the day liked to call the 'new illuminant'.[1]

The Electric Lighting Act of 1882 had, in a sense, thrown down a gauntlet to municipalities throughout the country. They could not ban electricity, as many of those that ran their own gas works would have liked. If they did not apply to run their own electricity scheme, that would allow in a commercial company which would seek to profit from a local monopoly. If they set up a scheme themselves, councils would have to pay for the installation, but any profits could be returned to the ratepayers. If they decided to allow in a commercial company, it did not cost them anything but the profits would go to the shareholders. Faced with this dilemma, many councils like Leeds would apply for an order that would allow them to introduce a scheme and then do nothing about it. This dog-in-the-manger approach could be drawn out over a number of years.

Compared with Leeds, the neighbouring Yorkshire town of Bradford was a little bit quicker off the mark. It took only six years to make up its mind that it would be best if it set up and ran the new electricity system itself. It was the first town in the country to do so, and it was an inspiration to many others who chose the 'municipal trading' option under the terms of the electric lighting laws. Bradford had first applied to the Board of Trade for a provisional order to run its own scheme in 1883. The municipal lights were finally switched on in 1889. It was gas that got in the way of electricity. Whereas towns like Godalming and Chesterfield had welcomed the competition electric lighting posed for their local privately run gas companies, Leeds and Bradford both owned and ran their own gas works. Prominent ratepayers were investors in these companies, which paid a good dividend. Despite the scares of Thomas Edison's premature predictions about the demise of gas lighting, the occasional show of arc lamps presented no real threat to the established order. A little more worrying was the gradual up-take of the new electric lighting by various industrial works, which installed their own generators on private property and were therefore free from the restrictions of the lighting Acts. They were customers lost to the municipal gas works.

While Leeds established a separate electric light committee early on,

Bradford chose to leave its existing gas committee with responsibility for decisions about the introduction of electricity. Both councils spent many hours over several years debating the pros and cons of, first, electric lighting per se, and, second, whether it was best to call in a private company to supply the 'new illuminant'. For some time both councils took the view that the new technology was not yet tried and tested enough to invest in, except for special installations such as the lighting of a new town hall. This was reasonable enough: the technology was changing rapidly and improving visibly. Manufacture of incandescent lamps was greatly expanded, especially after the merger of Swan and Edison as Ediswan, which put an end to the patent dispute between them. Generators were improving, as was all the equipment, transformers, cables, switches and so on.

Like a great many other local councils, Bradford and Leeds used the application to the Board of Trade to run their own schemes to fend off commercial companies. However, until the law was amended in 1888 in favour of commercial companies, there was not much pressure to do anything. Then, in 1889, with the possibility of a franchise lasting forty-two years rather than just twenty-one, the dormant electricity companies sprang into action. The pressure was on councils like Bradford and Leeds to make a decision about whether or not they wanted to run a scheme themselves. It was then that Bradford stole a march on hapless Leeds.

On 20 September 1889 the *Leeds Mercury* reported the formal opening of Bradford's electric light installation. The Corporation had borrowed about £25,000 to pay for the work, which was overseen by its Gas Supply Committee. Pointing out that it had been six years since the council had first applied for permission to introduce the scheme, the *Mercury* commented: 'It will thus be apparent that the Corporation have not shown much haste in the matter, one consideration which would necessarily have great weight being that it was not desirable to do anything which might interfere with the financial affairs of the gas department, in which more than half a million of ratepayers' money is invested and from which an annual profit of about £40,000 is received in relief of rates.'

However, the Gas Committee of the Corporation had noticed that more and more gas was being used, especially to power small engines and for cooking, and it did not look as if electric light would undermine its profitability as they once feared. That was reassuring, but the time had come anyway when Bradford Corporation either went ahead with its own scheme or allowed in one of the commercial companies knocking on the door. 'Not being able to allow others to meet a general demand,' said the *Mercury*, 'the Corporation could not with a very good countenance persist in a policy of inaction towards the new illuminant.'

Bradford had used its delaying tactics to good effect, taking time to consult electrical engineers who advised on the design of the installation. It had set up a power station with steam engines and Siemens generators capable of supplying 5,000 incandescent lamps, each of sixteen candlepower. The *Mercury* report continues: 'The present intention of the Corporation is not to supersede the use of gas for lighting the streets but to deliver the electric current to certain points in the centre of town, whence owners or occupiers of premises may, at their own expense, arrange for its being connected with the establishments.'

By the 1890s there was no longer an expectation that electric lighting was essentially an alternative to gas lighting in the streets. Arc lights were too expensive and incandescent lamps were not really suitable. Nor were the lights principally for the residents of Bradford. 'The cables are laid in trenches underground on each side of the street,' reported the *Mercury*, 'but at present only one-half of the central part of the town will have the advantage of the arrangement, namely, Market-Street, Kirkgate, Darley-street and intervening thoroughfares. The apparatus is to be worked on the low-pressure system, thus avoiding risk of fire or injury to persons. A large number of tradesmen have made arrangements to have electricity "turned on" in their shops, the applications, in fact, being greater than the committee anticipated.'

A rival newspaper to the *Leeds Mercury*, the *Yorkshire Post*, was dismissive of the Bradford scheme, taking the avant-garde view that AC current would have been better because it would have enabled power to be sent longer distances and would have kept the electric cables much neater. Bradford Corporation, however, was unruffled. It made a loss on electricity of £1,200 in its first year but extended power supplies rapidly, and by 1896 it was making a profit of £3,500 and planning a new power station. By the turn of the century it was an enthusiastic promoter of electrical goods, hiring out electric motors and water heaters. One novelty was a 'copper bronchitis kettle' which cost £3.15s.: a good deal less than a doctor's bill, according to the council's propaganda.

Just after the Bradford scheme began operation, a letter was published in the *Leeds Mercury* castigating Leeds County Council for its failure to come to any decision about electric lighting. In the issue of 16 November 1889 a correspondent, using the nom-de-plume 'Lux', noted that the Town Clerk felt that the council needed to 'protect itself' from overtures being made by companies that wanted to bring electricity to Leeds:

> Was there ever a more ludicrous, a more disgraceful situation? There are here in Leeds thousands, nay, tens of thousands of men and women whose lives are made daily more miserable, and whose deaths are hastened by their working in gas-fumed manufactories, and yet the offer of the electric light is to be petitioned against in order that 'the Corporation may protect itself'!
>
> By all means let the Corporation take steps to protect itself. Dark corners ever dread the broom: but the ratepayers should consider whether they like their money to be spent in such protective measures. What is the meaning of such a statement by a responsible official? Why, in the name of common sense, should the threatened approach of these three companies (in most healthy

> competition apparently, and each anxious to spread the light) alarm the serenity of the Town Clerk and disturb the sweet harmony of the Council? If the Corporation, by permission of the Legislature, are unfortunately entrusted with a monopoly of electric lighting in Leeds, are they to be permitted for ever not only to display their own want of enterprise but to hinder that of others?

Leeds hung on for another year without making a decision about a lighting scheme. In October, with time running out before the Board of Trade blew the whistle and appointed a private company, the chairman of the Electric Lighting Committee, a Mr Hardwicke, proposed to the full council that they apply for an order to undertake the installation themselves. The motion was seconded by a Mr Willey. Having proposed council control, both Mr Hardwicke and Mr Willey spoke *against* the motion. They thought it would be better to have a private company, which would be more dynamic and which would cast its net wider, and, if everything went wrong, the bill would not be laid at the foot of the ratepayers.

Through the years of machination in Leeds, the anonymous columnist of the Saturday edition of the *Leeds Mercury*, who wrote under the name 'Jackdaw', had poured scorn on the council. Now Jackdaw was apoplectic:

> Liberty of opinion is every man's possession but when the chairman of a Committee gets up and proposes something in which he avows that he has no faith, and another member seconds the proposition in terms only perhaps a little more farcical than the first, then we must observe that the public are not being fairly dealt with. Local bodies exist to *limit as far as possible private monopoly in the interests of the community* [my italics], but our local administrators have interpreted their duty otherwise, and now the electric light is to be provided by private enterprise. The Council cannot undertake the risk, forsooth! In its wisdom it is going to wait until the monopoly

is worth a heavy price, when, of course, it may judiciously be purchased.

Though the 1888 amendment to the Lighting Act had given commercial companies forty-two years' grace before they could be bought by a local authority, there were many agreements which allowed this takeover by the ratepayers to happen much sooner. In the case of Leeds, the agreement with the successful bidder for the lighting installation gave a number of options which made it possible for the council to buy it out within ten years. As 'Jackdaw' foresaw, a price would have to be paid in the form of 5 per cent dividends to the original shareholders.

It was the charmingly named House-to-House Electric Light Supply Co. Ltd that won the contest to provide Leeds with the new illuminant. Robert Hammond, formerly of the Brush Company, the pioneer scheme for street lighting in Chesterfield and business partner with Ferranti, had established this new company in 1888. The scheme he had started in Brighton, on the south coast, had prospered when all others had failed, and Hammond had fifteen miles of cables feeding lights there by the end of 1887. He had had his fingers burned acquiring patents, first of Lane Fox's incandescent lamps, and later of Ferranti's generator, and all he proposed to do in Leeds was to install a lighting scheme using the best equipment available.

The first House-to-House scheme had been in a fashionable part of London, at West Brompton, and was successful early on, supplying 248 grand houses from nine miles of mains early in 1890. A feature of the Brompton scheme was its very efficient central station, and the company had ambitions to sell this expertise both at home and abroad. The steam engines were provided by the Leeds firm of J. Fowler, who, in 1891, went into business as a specialist builder of local power stations with House-to-House in a venture named the Leeds and London Electrical Engineering Co. Ltd. This company applied to the Board of Trade to supply twenty districts in London and more than 200 in the rest of the country, with subsidiaries set up in different regions.[2] The one chosen to light up Leeds was the Yorkshire House-to-House Electricity Co., which got Board of Trade approval at the beginning of July 1891.

Whereas ten years earlier street lighting had been put up in Godalming in a few weeks, the process of establishing a commercial power station was now a relatively large financial and technological undertaking. Yorkshire House-to-House had to raise money from shareholders, buy a site for the power station, advertise for customers and comply with Board of Trade regulations – more than eighty of them, detailing the price that could be charged, the accuracy of meters, precautions for not interfering with other services and so on. The take-up of shares was encouraging, with most of the investors local businessmen or professionals. In that sense it was a local company, though it had begun as an offshoot of Hammond's Brompton scheme. Whereas Bradford had played safe with a low-pressure DC system, Leeds got approval for an AC scheme. At first each customer would have their own transformer to step down the current coming into their premises. Later there would be sub-stations to do this more conveniently. The expectation was that once the electric light had been proven, demand would grow rapidly. Again, the early signs were promising. Shops and businesses began to apply for a supply of electricity towards the end of 1892, when the power station and laying of cables was due to be complete.

House-to-House had an elaborate discount scheme based on the hours of use of their supply; the greater the consumption the lower the price. But this was of no interest to the very large working-class population of the town who lived in 'two-up, two-down' terraced houses. In his detailed account of the Leeds scheme, *An Early History of Electricity Supply*, J. D. Poulter pointed out that a typical tenant of one of these houses, such as a tram driver, earning under 22 shillings for a sixty-hour week, was unlikely to pay the minimum 33 shillings a quarter for House-to-House electricity.

Much of Leeds would remain gas-lit, and candles would still burn in many windows for years to come. It was the owners of the big houses in the fashionable suburbs of Leeds, and the shops in the centre of town, which were the main customers of House-to-House. By the end of 1892 there were forty-six contracts signed for a total of just under 3,000 lamps, with an average of sixty-four lamps per contract. A year later there were 139 customers, provided with a total of more than 130,000 lamps. Electric lighting was clearly hugely popular with shopkeepers and the well-to-do. It was also installed in most of the institutions in town, including the Great Northern Railway station and goods yards, the General Post Office, the General Infirmary, the Co-operative Society's new building, the various political clubs and the Yorkshire Penny Bank. The Great Synagogue was another customer. Leeds Parish Church had electric light installed, but kept some gas jets in place just in case.

Year by year the Leeds scheme was extended, and Yorkshire House-to-House made good profits and paid dividends even though the company had to plough back a good deal of its earnings into new and more powerful generators. It was not long before Leeds County Council felt it was time this successful capitalist venture was taken into public ownership. This required a long period of negotiation and arbitration to fix the price the council would be obliged to pay to the shareholders. By 15 December 1898 all was agreed, and the hand-over took place at the headquarters of Yorkshire House-to-House. Shortly afterwards the company went into liquidation.

The founder of House-to-House, Robert Hammond, was not at all put out by the council takeover, as he and all the other shareholders did well financially. In fact in a pamphlet entitled *Municipal Electricity Supply* Hammond wrote: 'It is ... manifest that the consumers are better off if supplied by the Local Authority than by a company, because the price charged by a company must cover the cost of administration by directors and also a provision of a good dividend to shareholders, whereas in the case of supply by a Local Authority there are no directors to pay and no fat dividends to provide.' That opinion was expressed in 1893, when, as he showed in the pamphlet, most of the money invested in electricity works was private. But times were already changing and Hammond noted that most of the new schemes being proposed were to be council-run.[3]

Up to the outbreak of war in 1914, something like two-thirds of all public electricity supply was provided by local councils. Nervousness about investing in the new technology, which had delayed the decision at Leeds for so long, gave way to an enthusiasm for 'municipal trading' once the near certainty of making money had been demonstrated. But none of these schemes provided the whole of a town with electric lighting, which remained a luxury for householders. As late as 1910, a survey of inland towns in Britain estimated that only about 6 to 7 per cent of the population had electric light in their homes. In terms of 'light hours', gas was still far and away the most widely used form of illumination: a figure calculated for 1904 showed a total 28,000 million hours for gas as against only 6,000 million for electricity.[4] This was despite the fact that the price of electricity had fallen significantly and was competing with gas.

However, an innovation in gas lighting had given it a new lease of life: the gas mantle. When these little gauze covers for gas flames were first advertised in Britain in 1890, they were described as 'Incandescent Gas Lights', often with the slogan 'Electric Light Surpassed'. They had been invented by an Austrian chemist, Carl Auer, in the early 1870s, when he discovered by accident that certain oxides when heated by a gas flame glowed brilliantly. It took some years to find a formula that was commercially viable, so that the nineteenth century was nearly over when the little mantles became widely used in Britain.[5] The fact that the mantle was invented later than the filament light bulb appears at first sight to be anachronistic, but it is a reminder that the gas jets that heated homes and halls up to the 1890s were just naked flames of one sort or another. Mantles saved on gas, were easily adjustable, and gave out a very pleasant light.

Had the use of power station electricity been confined to lighting it would have given rise to a fairly small-scale industry. Rookes Crompton, and some others, had tried to diversify with electric cookers and irons, and enterprising towns like Bradford tried to encourage greater use of electricity by hiring out equipment. But, as William Siemens had prophesied, gas consumption would go on rising, as it was ideal for heating and cooking. The way forward for electric power would have to be well away from competition with gas. Thomas Edison had, of course, already indicated the direction electricity should take when he visited Deptford: get rid of those horrible steam trains and replace them with American-style electrified street railways. The modernizing of Victorian city transport in the United States and Britain was not long in coming, but strangely Edison himself had little to do with it. He remained preoccupied with lighting when several of his fellow countrymen took the lead in revolutionizing transport first in America and shortly afterwards in Britain. In the eighteenth century the first steam engines at work in America had been made in Britain and shipped across the Atlantic. In the late nineteenth century the traffic was in the other direction.

CHAPTER 8

THE AMERICANS RIDE IN

'Everyone who has experienced their comforts testifies to their value as a means of locomotion, excelling horse and steam trams in swiftness and smoothness, and being beyond comparison with both their elegance of aspect and brilliant electric illumination in the evening. Those advantages, it is said, make people forget all about the vista of iron poles and the long lines of crossing wires.' With these words the *Leeds Mercury* summed up in 1892 the advantages and drawbacks of the electric tram, recently arrived on a stretch of road in the town from America. Dilatory when it came to adopting electric lighting, Leeds was out in front when it came to introducing electric transport.[1]

There was just a little smugness in the newspaper report:

> The Electric tram-cars on the Roundhay-road are no longer the novelty they were, and the people of Leeds may reasonably be called

> upon to contemplate the extension of the system without having it said that they are being brought face to face with a startling new idea... As is generally known, the Thomson-Houston system, distinguished from others by its overhead wires and posts – a serious objection, in the opinion of many – is very popular in American towns, and in Boston, especially, it is largely adopted... The Roundhay-road is the only thoroughfare in England where such electric trams can be seen.

As in every other town of any size in Britain, the most widely used method of public transport in the late nineteenth century was the horse-drawn tram. Though there had been horse-drawn railways in the mining areas of Britain since the eighteenth century, and one or two passenger services, before the steam locomotive took over, the idea of having 'street railways' in towns was American. The first horse-drawn streetcar appeared on the roads of New York in 1832 and they rapidly spread to other cities, where they provided a relatively cheap form of transport. It was a while before the urban horse-drawn railway crossed the Atlantic, and when it did so it was brought by a wonderfully flamboyant, larger-than-life American called George Francis Train.

Train was born in Boston in 1829 and moved with his parents to New Orleans, where the family was struck down in a yellow fever epidemic. George was the only survivor, orphaned at the age of four.[2] He was raised by his mother's parents in Massachusetts until he went to work for a cousin of his father who had a successful business running packet boats to Liverpool. When he was twenty-one George was posted to the English port to oversee the business there, arriving in 1850. Shortly afterwards he went into business on his own, married and headed for Australia, where he made a fortune in the gold rush town of Melbourne. He then returned to Liverpool and, for reasons that have remained obscure, began to promote street railways in England. The first successful venture was in Birkenhead, across the Mersey from Liverpool, and Train moved down to London to try his luck in the West End. He got permission in 1861 to lay his tracks along the Bayswater Road to Marble Arch, but the chorus of upper-class protest put a swift end to the experiment. Apart from the general inconvenience of having a railway among horse-drawn cabs and carriages, Train's tram rails protruded from the surface of the road and were a serious hazard.

There was a hiatus after Train's retreat, but the idea had caught on, while the American himself appears to have suffered a mental breakdown and bankruptcy on his return to America. The Paris-based Compagnie Générale des Omnibus de Londres, which, for a time, had bought out all the London horse buses, tried to get a tram line going in the capital but was beaten back by opposition from a variety of interests. While it was acknowledged that a carriage on rails was a more efficient vehicle than a bus because there was less friction on the wheels, enabling a horse to draw a much greater weight, the laying of rails on English roads was a huge problem.

If you wanted to build a tram line you had to get permission through an Act of Parliament, which was expensive and time-consuming and gave opponents of the line a chance to raise every objection they could think of. It was precisely the same procedure that had been followed with the cutting of canals in the eighteenth century and the construction of steam railways in the nineteenth. When there was a clamour for the building of horse tramways in the late 1860s, Parliament debated the issue and passed a catch-all Act that laid down the rules for their construction.

The Tramways Act 1870 made tramway building a frustrating business, with many regulations designed to protect the public and the owners of private property. To ensure the rules were not broken, the building and running of tramways was to be overseen by the Board of Trade, the same department which was later in charge of electrification. Although in most respects the Tramways Act followed the precedent of canals and railways, there was one entirely novel clause, described by the politician

Vesey Knox, in a paper to the British Association in 1901, as 'the most disastrous legislative experiment which has been attempted in England during the last half century'. The offending clause stated that after twenty-one years from the grant of permission to run a horse tramway, the local authority, or authorities, through which it ran, could buy it. There was to be no appeal and the price paid was to be the 'scrap value' of the tramway. This was the first instance of the clause that later bedevilled the Electric Lighting Act.[3]

Despite this clause, commercial tramway companies sprang up everywhere in the 1870s. The routes they took and the areas they served did not amount to anything like a sensible transport system but reflected the degree of opposition the promoters encountered. As the cheapest and most popular means of getting around the major towns, they were generally regarded as a 'working-class' form of transport. This ensured that they were excluded from the more salubrious suburbs: central London did not let them in. Local authorities showed little interest in owning or running horse trams, and there was no expectation that they would be too keen to buy them up when the twenty-one years of their lease were up. This was to change radically with the coming of the electric tram, which arrived just as the twenty-one-year franchises for the horse tram were expiring.

The German firm of Siemens & Halske had demonstrated an electric tram in 1879, and ran one with an overhead trolley wire at the Paris Exhibition in 1881. There were some early ventures in Britain too. In 1883 a short line was opened on the promenade at Brighton as a tourist attraction, and in the same year a tram driven with hydro-electric power ran in Antrim in the north of Ireland to take sightseers to view the geological wonder of the Giant's Causeway. Two years later an electric tram ran along the seafront at Blackpool, and another line was opened in Ireland between the port of Newry and the linen weaving village of Bessbrook. None of these early lines, however, replaced an existing tramway and they were not much more than novelties.

A more significant development was the trial of an electric tram in London, reported in the *Daily News* of 9 January 1890:

> We glide out of the station noiselessly, away into the high road. The car can run at a speed of twenty miles an hour; but of course that is not permitted. The facility with which it can be manipulated is surprising. In an instant, it can be made to crawl along at a snail's pace, or go away at full speed. Its driver once drove a 'bus'. He much prefers electricity to equine muscle. And he learnt how to control his electric car without any trouble in two or three days time. On this section of the tramway there is a very stiff gradient. The electric car goes up it with the greatest ease. In going down the incline the 'current' is turned off; the car goes by its own momentum, so that 'energy' is saved; whereas, even with the drag on, horses would have had to run all the same, thus expending their energy uselessly. *Each electric car has displaced sixty horses...*

The experimental tram described by the *Daily News* reporter ran on a section of the extensive network of tramways of London's North Metropolitan company. Its motor was designed by Moritz Immisch, a German watchmaker who had settled in London and become interested in the novel technology of electric gadgets. The Immisch motor was successful, but it was powered by a system that did not live up to its promise. The *Daily News* reporter described the 'electric stables' where the cars were charged up:

> We have visited the Electric Car premises at Greengate. They are a transformed mews. In place of the horses there are electric batteries; muscular 'energy' has given place to electric 'current'; the 'dynamo' has superseded the box of oats and chopped hay; the 'cell' takes on its 'charge' where erstwhile the friend of man crunched his grain; and, instead of the tight-breeched stableman with his sleeves

> up and his braces about his hips, and his familiar talk of ''osses', there is the electrical foreman, who is strong on 'volts' and 'amperes', who speaks mathematics, who looks to the 'insulation' of his wires...

The battery-operated trams that North Metropolitan had chosen appeared to be ideal replacements for the horse-drawn carriages. Each time the tram returned to the depot the batteries were changed, newly charged ones placed under the seats of the carriages and connected to the motor. No external wires were needed, nor any system for transmitting electricity to the tram, as it moved along tracks that had already been laid and did not need to be altered. But these trams proved to be a failure wherever they were tried, in both America and Britain. In a short time batteries shaken up by the movement of the tram had to be repaired or replaced, and they were not cheap to manufacture. One experiment in Croydon in 1891 ended in a messy disaster when acid from the batteries spilled out on to the passengers. The experimental tram that had seemed such a brilliant invention was abandoned by North Metropolitan in 1892, and horses were led back into the East End stables. Immisch fitted his tram motors into boats, and, on the upper Thames, electric canoes were all the rage for the gay young things of late Victorian and Edwardian England.

Meanwhile, in America, after years of experimentation, one or two determined engineers had managed to design really economic tramway systems. In the forefront was Frank Sprague, a graduate of the United States Naval Academy, a man of great inventive genius who deserves to be better known. He was born in Milford, Connecticut, in 1857, the son of a hat maker.[4] Taken under the wing of an aunt when his mother died, he showed enough promise to be put in for the Naval Academy examination, for which there was stiff competition. The standard of science teaching there was high, and Sprague gained an interest in electrical apparatus, such as Gramme's dynamo, which was causing a great deal of excitement. He graduated in 1878, and, while an ensign at sea, filled notebooks with ideas of electrical gadgets of various kinds.

After two years away, he returned to Newport Naval Station where he was able to conduct some experiments. He wanted to make it to the Paris Exhibition of 1881 but was too late. However, he was able to visit the similar event staged at London's Crystal Palace the following year when he crossed the Atlantic with the Mediterranean Squadron. This was a turning point in his life. He was appointed secretary to the jury giving prizes at the Exhibition and met Edward Johnson, Edison's representative in Europe.

While he was in London, Sprague took a few trips on the smoke-filled Metropolitan and District underground, which turned his thoughts to the possibilities of cleaner electric transport. He discussed this and other electrical matters with Johnson, who was impressed enough with the twenty-five-year-old Sprague to suggest he apply for a job with Edison at Menlo Park. Edison took him on in May 1883, and sent him to install a new kind of electric wiring system at a central station in Brockton, Massachusetts. Sprague stayed there as the operating engineer, which gave him time to work on his ideas for an electric motor. Edison, despite his later boast when he visited Deptford in 1889 that he had offered to electrify London's steam-driven underground, really took little interest in railways. He did experiment with an electric line at Menlo Park, but it was not a commercial success and he put all his efforts into lighting.

Sprague left Edison in 1884 to develop his motor, and, with Johnson, set up in partnership the Sprague Electric Railway and Motor Company. While Johnson bankrolled him, Sprague worked on the fine-tuning of his motors, which were manufactured by Edison's Machine Works. An agency sold them across America wherever they could find customers. Often they were installed in factories that already took Edison electric light: the same generators that lit the factories at night could work the machines in daylight. Several Sprague motors were used to power lifts, one a freight elevator in a six-storey building in Boston. Sprague's business partner, Johnson, was president of the Boston Edison Company and was successful in promoting the motors in that town, so that by 1887 seventy-three Sprague motors were in operation there.

With sales of his motors growing – about 250 were in use in the United States by 1887 – Sprague set up his own factory in New York to experiment further, while Edison Machine Works turned out his standard and proven models. Though he was developing motors for all kinds of uses, Sprague began to concentrate on the problems of electric traction, and in particular on a project to convert the steam-driven elevated railroads of New York. Electrifying a steam railway should have been, in many respects, more straightforward than tackling the horse-drawn trams of the city: the railroads were segregated from the streets, so there was no danger of electrocuting pedestrians with a power line, and much of the existing infrastructure could be used. There was, too, an urgent need to upgrade the steam lines, which could not cope with existing traffic.

Sprague made some experiments on a section of the 'L' in 1885 at Twenty-Ninth Street. Later, in 1886, with an extended area of track, he was able to demonstrate electric cars running smoothly and with faster acceleration than the steam trains. No heavy locomotives were needed to pull an increased number of passengers, as the carriages were, in effect, self-propelled. But Sprague failed to persuade the company that ran the Manhattan Elevated Railway Company that electrification would be of benefit to them. This might have been because the lead financier, Jay Gould, had been startled by the blowing of a fuse when he rode on one of the experimental lines.

After two years Sprague gave up on Manhattan and was soon to engage on a project which he knew would make or break him. A New York financier, interested in the development of Richmond, Virginia, approached Sprague with a proposal to build an entirely new electric line. It would link the town with a factory on a route of steep hills, soft ground and sharp curves. The line would be twelve miles long and would carry forty cars. Horse and cable power were out of the question. Yet Sprague had only a sketchy notion of how he would build this line. Nevertheless, he signed a contract in May 1887, committed to completing the line ninety days after the track had been laid, and with the promise of $110,000 in cash if it had operated satisfactorily for sixty days after it began to run.[5]

Everything that could go wrong did so. The track sank into the soft soil in many places, the motors burned out with the stress of hauling the cars uphill, and new gearing had to be devised. Designing a mechanism for picking up the current from an overhead wire proved difficult, and there were many problems with insulation and the working of switches. In July Sprague was bedridden with typhoid fever. Test runs on the track made in November dashed all hope of completing electrification in ninety days, and the final payment was reduced. In February 1888 the line went into service, but it was fraught with problems. The cars kept coming off the rails and the motormen (drivers) had to carry ladders to reconnect the cars to the overhead trolley poles, as well as a large piece of timber to lever the cars back on to the rails.

Sprague wrote later that if he had known when he began what the problems would be, he would not have taken on the Richmond line. He spent almost twice the money he was paid before the line was working satisfactorily, losing $70,000. But he had solved all the basic problems of running a sizeable tram system. He survived financially because his motors for industry and elevators were profitable, and he was soon reaping the benefits of the Richmond scheme when it was realized that it was 40 per cent cheaper to run than a horse tram service. It was not long before Sprague had other contracts and electric streetcars began to replace horse carriages across America. At the same time, entirely new electric streetcar lines appeared, often promoted by developers who bought cheap land on the outskirts of the rapidly expanding cities and hoped to profit when the streetcar brought about suburban building.

This gave rise to a huge market in America for electric streetcar equipment.

The main streetcar rival to Sprague was the firm of Thomson-Houston. The two partners, Elihu Thomson and Edwin Houston, had got to know each other when they taught at the Central High School in Philadelphia. Born in Manchester, England, in 1853, Thomson had arrived in America with his parents when he was just five years old. He became a pupil at the Central High School and was asked to stay on as a chemistry teacher when he left. Edwin Houston was a senior teacher there and the two conducted some experiments together. One collaboration was in order to debunk a claim by Edison that he had discovered an 'etheric force', which produced sparks, but was not electric in nature. The new force turned out to be electro-magnetic waves, which were later to be generated as wireless signals.

Thomson became interested in dynamos and wrote several papers on their operation, which attracted the attention of the Franklin Institute at a time when it was holding a competition to discover which dynamo model it should buy. Charles Brush was the successful contender, and Thomson got to know him and began to take an interest in arc lighting. In time Thomson and Houston set up together to make and market arc lighting equipment, which they did with considerable success over a number of years between 1880 and 1887. They were given backing by a group of Boston businessmen who funded the Thomson-Houston company from 1883 with a managing director named C. A. Coffin, a salesman who knew nothing about electricity himself.

It was Coffin who, in 1887, persuaded Thomson to branch out into electric railways and to buy up the patents of one of the pioneers, Charles J. Van Depoele, a Belgian cabinetmaker. Van Depoele had emigrated to the United States in 1869 at the age of twenty-three to seek his fortune making furniture in Detroit. His interest in electric lighting and power for railways began almost as a hobby. He invented his own trolley-pole method of feeding current to streetcars and built a number of lines before Thomson-Houston bought him out in 1888.

There were other makers and promoters of electric streetcars and railways in America around 1890, but Sprague and Thomson-Houston were by far the most successful. It was then that they began to take an interest in the potential market in Britain. They had discovered that by far the most economic way to run electric trams was with overhead wires feeding current to an articulated trolley pole. Though the wires were unsightly, the comfort and speed of the electric trams was more than adequate compensation in most places. And it was not long before American streetcars appeared in British towns, to the chagrin of those, like Vesey Knox, who witnessed the eclipse of domestic industry with barely controlled rage:

> Electricity can neither be profitably generated in a moving vehicle nor economically stored there, but means were found, by the overhead wire or the conduit, of supplying it to the tramcar as it went along. This immediately reduced the cost of locomotion by half, while also increasing the speed and pleasantness of street transit to a degree never before possible in a public vehicle. Seventy passengers can be carried a mile for sixpence in working cost, and they can be carried to the precise spot to which they want to go.
>
> There is no country in the world where this discovery should have had a readier welcome than in England. When it was made, England was still the greatest steel-producing and engineering country. It could still claim the largest experience in the best railway construction, other than street railways. In no other country was there so large a contiguous urban population as in London, Lancashire and the West Riding... In ordinary course England should have been the pioneer in the construction of electric tramways... It is needless to say this has not been the course of events.[6]

And so it was that Leeds, in the heart of what had been the greatest industrial nation in the world, enjoyed the fruits of American technology with its Thomson-Houston experimental line, opened late in 1891. Typically it took another four years before Leeds decided to go over to electric trams entirely, and in doing so it spurned the offer of its House-to-House suppliers of electric lighting to generate its tram power. As the electric tram spread, a pattern emerged in which local councils exercised their right to take over the trams and shortly after electrified them. In most cases a power station separate from that providing electric lighting was built to power the trams. This ensured that the electric supply industry, even at a very local level, was fragmented and therefore relatively inefficient.

However, the electric tram that Sprague and others had created was hugely successful. Overhead wires and the trolley pole were accepted in most northern cities and in the Victorian suburbs of London. A more expensive conduit system with the power rail at road level was insisted upon in the inner London suburbs. Whereas the horse trams had been commercial, the electric trams were mostly municipally owned: it was calculated that in the early 1900s about 80 per cent of electric tram passengers in Britain rode on council-owned services. Often the council did not run the tram lines it owned, though some did.

There was tremendous local pride in electric trams, which became such a feature of cities from the early 1900s until their rapid disappearance after the 1950s. It is often said that a ride on an electric tram, which was especially romantic at night with its lights ablaze, was really the first experience of the wonders of the new technology for the majority of people. City trams were especially suited to the smoke-shrouded towns, for when the fog descended, and visibility was no more than a few feet, they could always find their way back to the depot, often leading the other road traffic. They were also remarkably safe, all of them fitted with a kind of human equivalent of the American railway 'cow-catcher' in case they ran into a pedestrian.[7]

It was a constant lament of the electric tram companies, however, that the speed restrictions imposed on them by the Board of Trade and local regulations put them at a disadvantage when they were in competition with motorbuses, which began to run in the early 1900s. In January 1905, the secretary of the Tramways and Light Railways Association, Ernest Benedict, wrote to the Board of Trade to point out that whereas a motorbus was permitted to travel in parts of London at fourteen miles an hour, trams were generally limited to twelve miles an hour even on the fastest stretches of their route. A bus driver could choose his own speed on tight corners and difficult stretches of road, whereas a tram driver was restrained by official speed limits at every turn, often limited to speeds of four miles an hour on a bend. Benedict argued that the speed restrictions, which the Board of Trade refused to reconsider, had 'hampered the tramway industry in Great Britain and Ireland, and has helped to render the return on capital less than in any other country'. However, while the restrictions on speed were irksome for the commercial tramway companies, the municipal operators, such as the London County Council, appear to have been unconcerned about the limits they set themselves. Though over much of London the electric tram was not allowed to run at more than ten miles an hour, this easily outpaced the horse-drawn tram and, in fact, was the average speed for traffic in the capital at the beginning of the twenty-first century.

In London, electric power did away not only with the thousands of horses that pulled the trams, it made possible an entirely new form of big city transport as well as the modernization of the old underground railway. The earliest underground lines were the smoke-filled Metropolitan, District and Circle lines, which both Sprague and Edison had ridden on during their visits to London. These lines could cope with steam-driven trains because they ran just below the surface of the road, the tunnels created by a 'cut and cover' technique that allowed some escape of smoke at intervals. Electric power made possible an entirely new kind of underground line that ran deep below the surface, where a steam train would have been quite impractical.

The technique of burrowing through the subsoil of London had been developed by Marc Brunel and his son Isambard in their epic creation of the Thames tunnel, which was opened for pedestrians in 1843, sixteen years after it was begun. Brunel's method of cutting through soft soil with a 'shield' was improved by the engineer Peter Barlow, whose 'great shield' was used to cut a tunnel from north of the Thames by the Tower to Rotherhithe in the south. A train powered by a pulley system ran through this tunnel carrying passengers for a few years. But it was not popular, as there were frequent breakdowns, and it closed in 1897. By that time the first electric underground line in London had opened: the City and South London, which ran from King William Street near the Monument, just north of London Bridge, to the suburb of Stockwell, south of the river.

The first plan for this line had been to use cable traction, which was popular then in some American cities, notably Chicago. However, when the line was nearing completion, the directors of the operating company took the bold decision to make it an electric underground train.[8] They were able to find British firms to provide the generating equipment and motorized carriages, and the line was opened with pomp and ceremony on 18 December 1890. There were four stations between the City and Stockwell, calling first at Borough, heading south, then Elephant and Castle, Kennington and Oval. The stations were gas-lit and had wooden platforms. Signalling was mechanical. There were teething troubles: the motors tended to burn out while carrying heavy loads. Siemens provided improved engines that were more reliable. The City and South London demonstrated not only to Londoners, but to the rest of the world, that it was possible to carry passengers on railways deep below ground. They were actually cheaper to construct than the old 'cut and cover' lines because they avoided a lot of the cost of compensation to property owners.

In the eyes of American engineers and financiers, London was a huge potential market for new transport systems, and when the next deep-level tube was proposed, from Shepherds Bush to the City, most of the technology and expertise came from across the Atlantic. What is now called the Central Line, running east to west through the heart of London, was built with international finance and American technology. It had a flat-rate fare of two pence and was soon dubbed the 'tupp'ny tube'. When it was opened in 1900, the American magazine *Science* was able to boast:

> The Central London Railway, the 'Electric Underground' of London, the 'two-penny tube' is one of the most important and, in some respects, far the most remarkable example of the work of the American electrician and engineer in Europe, perhaps in the world... It was found necessary to come to the United States to secure its exceptionally large and powerful machinery and motive power. It is, in fact, an American electric railway in operation in London, the center of the brains and business in London.[9]

The trains were built by the General Electric Company, the giant formed by Edison and his backers, which had gobbled up so many other companies. Thomson-Houston provided some of the electrical equipment, and the lifts that took passengers down to the platforms were powered by Sprague motors. The power was also provided by an American firm, E. P. Allis & Co., and the boilers to drive the steam engines that turned the generators were American, built by Babcock & Wilcox.

The success of the Central Line encouraged investment in more lines which had been proposed in the 1890s but lacked financial backing. The *Daily News*, which took a great deal of interest in London's transport, published in December 1900 a map of all the lines projected at that time.[10] It is quite remarkable that most of these projected lines were actually built, and in a very short space of time. In the same period the old steam-driven District and Metropolitan lines were electrified. This huge programme of electric railway building was driven with a mad enthusiasm by an American who, on a trip to London in 1900, announced that he would transform the capital's transport system just as he had the railways and tramways of Chicago. Under the headline 'A Man of the Moment', the *Daily News* of 2 October 1900 devoted nearly a whole page to Charles Tyson Yerkes:

> The announcement that Mr Charles T. Yerkes, the American Street Railway magnate, is to play a leading part in organising a more effective system of city passenger traffic than at present exists for London's teeming millions will be welcome news to those who have knowledge of what Mr Yerkes has accomplished in a similar direction in his own country. To his efforts more than those of any other man, the cities of Philadelphia and Chicago owe their emancipation from the thraldom and delay of horse-driven public conveyances. Not that Mr Yerkes has invented any new method of locomotion, but he was quick to perceive how important a part it was possible for steam and electricity to assume in accelerating the movements of the inhabitants of populous cities in their daily journeyings to and fro. He found vast multitudes struggling with antique methods of transit, and while others were thinking and dreaming on the subject, he tackled the problem, and quickly transformed the condition of street traffic from jogtrot, humdrum, horse propulsion to electric or steam power.

It is a mystery how the *Daily News* was persuaded that Yerkes was some kind of saviour of the American commuter. Even more puzzling is the fact that Yerkes persuaded Parliament that he, and not the millionaire banker J. P. Morgan, was the man to rid London of its smoky underground and remaining horse trams, and to cut new deep-level railways.

Perhaps the politicians had been influenced by the gushing nonsense printed about Yerkes in the *Daily News* report:

> He has never allowed himself to be drawn into public life, and the blandishments of society have small attraction for him. One hobby he has apart from business, and that is art, his picture gallery in New York being one of the finest in the country... Chicago has found in him one of its chief benefactors, one of his recent gifts to the University of Chicago taking the form of a telescope, designed to be the largest and finest in the world and to cost £100,000. Mr Yerkes is a man of robust constitution and abstemious habits, with a genial personality and a readiness of thought and action that stamp him as a man of business distinction.

In fact Yerkes was sixty-four years old and not in very good health when he began his campaign to transform London transport. A native of Philadelphia, he came from a Quaker background but early on demonstrated a greater interest in personal aggrandizement than philanthropy.[11] With a small inheritance from an uncle, he set up as a banker and quickly got into trouble. His stock-in-trade was manipulation of public funds, which, through bribery and threat, he managed to siphon off for his own speculations. Caught out in the financial panic of 1871 caused by the great Chicago fire, he and his accomplice, the Philadelphia town treasurer, were convicted of robbing the public purse. Yerkes was sent to the State Penitentiary for Eastern Pennsylvania, once notorious for its regime of solitary confinement, but appears to have survived the seven months he served of a two-year sentence without much hardship. Though he was married, and his wife visited him in jail, he was also cheered up by a ' prison angel', a beautiful young woman who is believed to have been one of his many lovers. Yerkes was pardoned, allegedly because he was in a position to blackmail the local politician who secured his release.

Yerkes left Philadelphia in 1880, and had many business ventures before he settled in Chicago and began to buy into, and buy up, the city's street railways. He brought in financial backing from Philadelphia and learned to run both horse-drawn and cable-drawn tramways at a profit, always with a swindler's sleight of hand. The companies were vastly overvalued on the stock exchange, swelling his apparent wealth, and companies were created which overcharged for their services. At first Yerkes did nothing to improve the streetcars, some of which ran without roofs in the bitter Midwest winters. He was ruthless with anyone who objected to a new tram line: he would get his men to put the posts and wires up during the night. While he became enormously wealthy, Chicago suffered. As Robert Forrey put it in his 'Charles Tyson Yerkes: Philadelphia-Born Robber Baron': 'Almost everyone in the City rued the day Yerkes had arrived.'[12]

Yerkes built himself lavish mansions in both Chicago and New York. In his Fifth Avenue second home he installed a bed that had allegedly belonged to the mad King Otto of Bavaria, the footboard of which was decorated with erotic motifs. Here Yerkes entertained his women friends. Divorced from his first wife, in 1881 he married Mary Adelaide, who was possibly the same woman who had been his 'prison angel' and was half his age. Beautiful, talented young women often caught his eye, and he showered them with gifts. One catch was a great-niece whom he ran into in New York: she was reputedly tucked up in the Otto bed with Yerkes when his wife made an unexpected visit. Yerkes's favourite companion, however, was Emilie Grigsby, daughter of a brothel-keeper in Cincinnati, on whom he lavished a mansion near his own on Fifth Avenue.

Yerkes had created a wonderful lifestyle, in which the fortune he made robbing Chicago he spent in New York. But it all began to fall apart when he attempted to bribe councillors to award him franchises for his trams and railways which would run for 100 years. When his subterfuge was made public, he was driven out of Chicago. He had often visited Europe, chiefly to collect art works for his mansion, and now fled to London with his fortune – and an extraordinary entourage. Yerkes put up at Claridge's with his second wife, though the two were estranged, and installed a seventeen-year-old girl, Gladys Unger, in a suite at the Hotel Cecil, where he had his rented offices. Shortly afterwards, Emilie Grigsby arrived to live in a rented apartment nearby.

The time was certainly ripe for someone to making a killing developing London transport, and the prospects looked good for Yerkes, though he was not in a position to bludgeon his way into big city franchises as he had done in Chicago. What he did, in effect, was to play a kind of Monopoly game with the existing proposals for new tube lines and the ailing companies running the Metropolitan and District steam railways. British investors showed little interest in London transport, but Yerkes was able to excite the enthusiasm of Americans. As he and his companies bought up existing and proposed lines they gained a firm hold on what became known as the underground system. The bulk of the investment was American and so was most of the equipment, which in a few years brought about the electrification of the District and Metropolitan lines, and the building of the first sections of three deep-level tubes, the Bakerloo, Piccadilly and Northern lines.

Work had just begun when, in 1905, Yerkes fell ill and returned to New York. His wife refused to allow him into his Fifth Avenue mansion and he stayed at the Waldorf-Astoria, cared for by Emilie, the brothel-keeper's daughter. When he was close to death his wife relented and called to see him, accompanied by her sister. When she saw his young companion she went wild and Emilie made a run for it. Yerkes died shortly afterwards and was buried in a mausoleum he had had built back in 1891 in Greenwood Cemetery, Brooklyn. In his will he left bequests to a hospital, and his fabulous art collection to the Metropolitan Museum in New York. It turned out he was more or less bankrupt.

However, Yerkes, and those he persuaded to invest in London transport, did leave a huge power station at Lots Road, Chelsea, which had been built to provide electric power for the District Line and was the largest of its kind in the world when it first went into operation in 1905. The siting on the Thames had the same logic as Ferranti's ill-fated Deptford station: a river wharf where the thousands of tons of coal needed to fire its boilers could be unloaded from freighters that came down the east coast direct from the mines of Northumberland and Scotland. The river provided ample supplies of the water that was needed as a coolant.

Electrification of the trams by the London County Council required a second power station, which was built at Greenwich, despite concerns that it might interfere with the instruments of the Observatory. Ferranti's Deptford station got a new lease of life when it was called on by the LCC to provide power while its own station was being built. The electrification of tramways and some railways – the main steam lines were not converted for many years – led to the building of larger power stations than those constructed for lighting. The Central Line had its power station at Wood Lane, and the Metropolitan a power station at Neasden. Although the technology was there to link them all up, they were operated independently in the early 1900s and often at the limit of their capacity. They were all reliant on steam engines, which provided the power to turn the generators. Just as there were differences of opinion about the merits of direct and alternating current, so there were disagreements about the best way to harness steam engines to generators, and whether it was best to run slow or fast. As power stations began to generate very high voltages, the traditional 'reciprocating' steam engines grew larger and more cumbersome and stimulated the search for a more efficient way of using steam to produce electricity. The solution was found not by an American, this time, but by the slightly eccentric and playful son of a distinguished aristocratic family with an estate in Ireland.

CHAPTER 9

A VERY BRITISH INVENTION

In January 1886, weeks of bitter weather froze lakes and ponds throughout Britain, from London's Regent's Park to the shipbuilding towns of the north-east. Everywhere revellers put on their skates and ventured on to the ice. In the Gateshead district of Newcastle upon Tyne they headed for Swan Pond on Sheriff Hill. Watching the fun, the Chief Constable, a Mr Elliot, had an inspiration. If Swan Pond could be lit at night, and the skaters paid a penny or so for admission, some funds could be raised for the local infirmary. The Pond was a long way from any of Newcastle's electricity supply stations, but Mr Elliot knew of a company that might be able to provide the lighting. The local firm of Clarke, Chapman & Co. had recently developed a new kind of portable generator which it called a 'turbo-dynamo', a compact piece of machinery mounted on two wheels and hauled by a single horse.

A memory of the event was published many years later in the *Newcastle*

Evening Chronicle: 'Elliot carted the turbine up to the ground, where it was set up. Lamps were hung around the pond and the turbine was got to work. I think Mr Swan, afterwards Sir Joseph Swan, provided the lamps. It was a great success from Elliot's point of view, because the place was so crowded that few people could really skate, but everybody paid to get in, to say they had skated by electric light. As far as I can remember, the frost lasted three days, and the Royal Infirmary benefited by about £100.'[1]

This was one of the first outings for the turbine, the prototype of a new kind of steam-powered generator which in a few years was to make all previous power station equipment redundant. It was the invention of an eccentric young man just in his thirties, who had bought into the Clarke, Chapman firm just as they were developing electrical equipment.[2] Raggedly dressed, often covered in oil, absentminded – he once forgot to take his young wife home after asking her to wait for him – Charles Algernon Parsons was from the Anglo-Irish aristocracy. His father was the Earl of Rosse, whose ancestral home, Birr Castle, was in the centre of Ireland, in what was then called Parsonstown. As the youngest son, Charles was not in line to inherit, but he enjoyed a stimulating childhood on an estate in which he and his brothers were free to play at mechanics. Their father was a distinguished amateur scientist and astronomer who encouraged an interest in all kinds of machinery, so much so that when Charles was just seven years old he helped his older brothers build a steam-powered car that could travel at seven miles per hour. Charles was the stoker. Sadly a cousin, Lady Bangor, fell from the steam car and died.

This was a very rich family that would tour in their own steam yacht, the *Titania*, in the summer and spend their holidays in the European alpine resorts. But Charles's father was no idle aristocrat. He was a skilled mechanic who designed and constructed his own telescopes for astronomical work. Charles clearly inherited that ability, making models of various machines from a young age and showing a determination to invent *something* useful, though he was not sure what it should be. He was educated by private tutors and was sent to Trinity College, Dublin, at the age of seventeen. His father had died when Charles was thirteen. From Dublin he went on to Cambridge, where he studied mathematics and was good enough to be judged 'eleventh wrangler' in the Tripos, wrangler being the term for those who achieved a first class degree. However, Charles himself said he spent more time rowing in Cambridge than reading, and he was never a theoretical scientist, taking little interest in mathematics in his later career.

In 1877, after he graduated from Cambridge, Charles signed on for a four-year apprenticeship with Armstrong's of Elswick, Newcastle upon Tyne, the firm founded and owned by William Armstrong, who would later electrify his house at Cragside. He served his four years and left with a letter commending his 'theoretical knowledge... constructive abilities and... promising business qualifications'. During his apprenticeship he had begun to experiment with electricity generation, and when he was just twenty-three years old he patented a steam engine that could drive a generator directly without the need for belts or pulleys.

When he left Armstrong's he joined a friend, William Cross, in Leeds, where for two years he worked on a system of rocket propulsion for torpedoes. During this time he met and, in 1883, married Katherine Bethell, who indulged his eccentricities – part of the honeymoon was spent watching early morning torpedo trials on a lake in Leeds. She contracted rheumatic fever that day and to aid her recovery the young couple went abroad, travelling around the United States, often in a horse and buggy, visiting New Mexico and California.

Back home, Parsons was able to buy himself into the Gateshead firm of Clarke, Chapman & Co., investing £20,000 to become a junior partner. They were not interested in torpedoes and asked him instead to develop their electrical department, in particular a neat and economic system for shipboard generators. The great liners built in British shipyards were important early customers for electric lights and machinery. Much of his experimentation was done in his own home, which he lit with arc lamps. He was forever inventing things, using whatever came to hand. To help his young wife he built her a steam-powered pram, using a biscuit tin as a boiler. It blew up.

More successful was his first attempt at building a new kind of steam-driven electricity generator. The model for this had been around for many years, and there were already some prototypes in operation, but they could not compete with established arrangements of the 'reciprocating' steam engine, in which the force from the pistons turned a wheel that was linked to the generator by a belt or rope. What Parsons worked on was the design of a steam turbine in which the flow of steam across a cylinder fitted with internal blades set it spinning. The big problem was getting the arrangement of the blades right, a difficulty that had defeated many other inventors. Once he had the turbine working it could be coupled directly to the generator without any pulley belt. Again there was a problem matching the speed of the spin of the turbine with the required spin of the generator.

By 1884 Parsons had a working model which is now recognized as the first 'turbo-generator' ever built. The technicalities involved in this extraordinary achievement are beyond description in lay language, but included a great deal of trial and error with the generator as well as with the turbine. After five years with Clarke, Chapman & Co., Parsons set up his own company, C. A. Parsons & Co., with a factory at Heaton, just outside Newcastle upon Tyne. He was finally in charge of his own manufacturing business. His eccentricities sometimes made him a difficult man to work with. An Australian whom he employed, Sir Claude Gibb, said of Parsons: '... he had Irish blood in his veins and was therefore lovable although often difficult. Possessing a considerable sense of humour and fond of telling anecdotes, he could be a fascinating and charming host, entirely considerate of his guests, but could also be quite unreasonably irritating and irascible.'

It was part of the agreement when he joined Clarke, Chapman & Co.

that the patents for Parsons's inventions would belong to the company. This led to a long dispute when he left them, which was resolved when they agreed to sell the patents back to him. In 1890 he sold the first of his experimental turbines to Forth Banks power station in Newcastle. Two years later the Cambridge Electric Supply Company, in the town where Parsons had attended university, bought the first sets of the fully developed turbo-generator. Improvements were continually made, and Parsons's company began to build larger and larger turbo-generators as their advantages were demonstrated. They were cheaper to run than the older style of steam-driven generators and much quieter, a great advantage in the early days, as public complaints about the noise and vibration of city centre power stations were common.

The greatest compliment paid to Parsons in these pioneer days was an order in 1900 from the city of Elberfeld in Germany for two of his turbo-alternators, at a time when Britain was heavily reliant, otherwise, on imported electrical technology. Again, in 1911, a set of Parsons's generators, which were at the time the largest in the world, was bought for Fisk Street power station in Chicago. By the outbreak of the 1914–18 war, turbo-generators were producing far more electricity than the old-fashioned steam engines and the revolution Charles Parsons had engineered was well under way. In his book *The Early Days of the Power Station Industry*, R. H. Parsons (no relation) wrote of the first turbo-generator: '… but for the invention of the steam turbine, the industry, as we know it today, could never have been created at all, for without the immense and efficient turbo-generating units, characteristic of modern power stations, electricity would have remained a costly luxury, instead of becoming one of the universal necessities of industrial and domestic welfare.'[3]

One of the factors that had defeated Sebastian Ferranti at Deptford was the cost of generating electricity on a large scale. As soon as the Parsons turbo-generator appeared, Ferranti recognized that his earlier generators were redundant and their production soon came to an end.

Parsons was undoubtedly a giant in the history of electricity, yet he remained a private figure with a horror of public speaking, very different from the flamboyant Thomas Edison, who was always ready with a quote.

A keen cyclist and an enthusiastic early motorist who serviced his own vehicles, Parsons was known for his endearing eccentricities. He once arrived late for a picnic after repairing his car with the springy ribs of the parlourmaid's stays, and he had a habit of lighting his pipe with a match struck on the milled surface of a gold coin. In his youth he rode to hounds (in pursuit of foxes), and he was keen on fishing and shooting. He treated invention almost as one of his pastimes. A rare quote from him, delivered in 1914 when he was given the freedom of Newcastle upon Tyne, captures this characteristic nicely:

> When I commenced to work on the steam turbine in 1884, with the hope of making it a practical success, it seemed to me, in spite of the fact that many had previously failed in their endeavours, that it was right in principle and that after a thorough experimental investigation, it should be possible to realise success. In short, I thought it was worth trying. Encouraging results followed, one improvement led to another and it gradually became an efficient motor. When it had beaten a compound engine driving a dynamo my old friend John Bell Simpson said to me one day when we were out shooting: 'Why not try it at driving a ship?' To which I replied that I thought the time was ripe for the attempt.[4]

It was not long before a new company was turning out Parsons's marine turbine engines and ships were reaching unprecedented speeds. The prototype, the *Turbinia*, with Parsons himself at the helm, made a spectacular unscheduled entry into the Naval Review at Spithead in 1897, racing among the Admiralty's fleet at 34 knots like some impish terrier. Just as he had revolutionized the way in which electricity was generated, Parsons's marine turbine engines soon proved their superiority at sea.

Although he was not from the north-east of England, that is where Parsons spent most of his adult life. It was a region that provided him not only with a great deal of sport – he stocked lakes with trout and had his own grouse moor – but with a vibrant industrial world of shipbuilding yards and engineering works. It was in this region that the next great stride was taken in the electrification of Britain.

CHAPTER 10

ELECTRICAL MESSIAHS

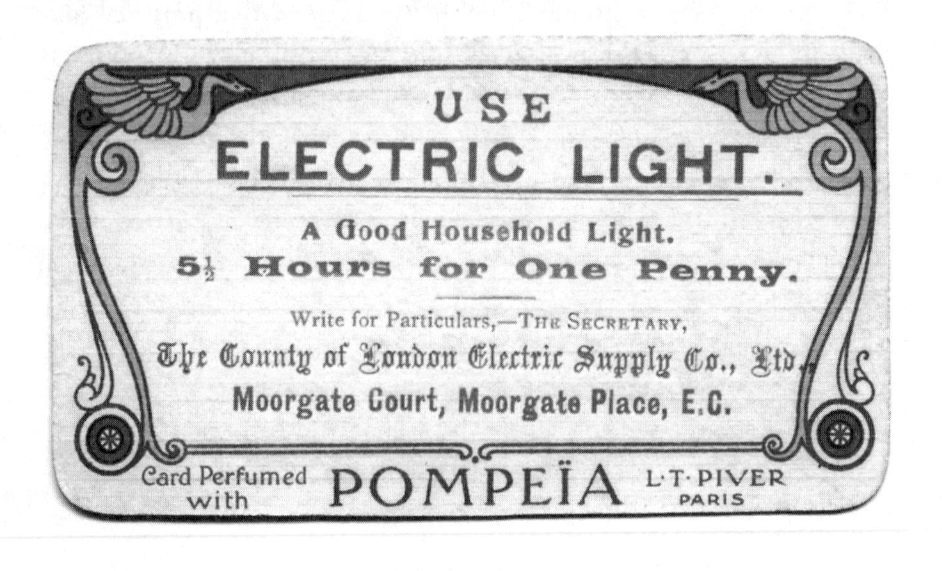

In March 1901, Charles Merz, then twenty-six years old, and his younger brother Norbert boarded the liner *Majestic* for New York, full of anticipation of their first visit to America. They were from a well-to-do family settled in Newcastle upon Tyne and could afford to pay £12 each for a first-class cabin they shared with a friend. But there the luxury ended. The *Majestic* ran into heavy Atlantic seas and they were both laid low. 'Norbert and I were both sick to begin with but soon recovered,' Charles wrote later in a slim volume of reminiscences. 'Marconi was on board going to initiate the first experiments for transatlantic wireless and he was dreadfully ill. I think this was the last time I succumbed to the sea on a steamship voyage, for which I was very thankful for I do not think I should have taken all the voyages I have taken if I had remained a bad sailor.'[1]

At the time he made this first visit to America, Merz had just set up in business with another young electrician, William McLellan. Over the two months that he travelled around America he sent McLellan his impressions in his scrawling hand, with sketches and diagrams. As Merz wrote later: 'The U.S.A. had not then embarked on that rapid electrical development but "the pieces were set for the game" and the next decade was to see wonderful progress.' He was impressed with the electric streetcars, or 'trolleys' as he called them, and some of the smaller power stations. Everywhere, American hospitality overwhelmed him. 'When an American firm get hold of you,' he wrote to McLellan, 'two dinners and lunches, a visit to a Factory and a weekend with them at the summer club are the least they suggest... However, by being positively rude I have managed to reduce this somewhat and hope to get off with a visit to a factory and a lunch.'[2]

Merz was no bon viveur. He was a very serious, dedicated and ambitious young man from a background which had instilled in him the virtues of hard work. He became, in the years immediately after his trip to America, a kind of electrical Messiah, helping to make his homeland, the north-east of England, far and away the most electrically driven region in Britain. At the same time, having overcome his dread of long sea voyages, he was a consultant on electricity schemes around the world, including a tramway in Melbourne, Australia, and railways in Argentina and India.

Neither he, nor his partner, William McLellan, were inventors like Swan and Parsons, nor were they manufacturers like Crompton and Ferranti. In 1899 they had set themselves up as 'electrical engineering consultants'. For their offices they converted a row of workmen's cottages in Newcastle and made it their business to develop an expertise in the most up-to-date equipment and systems. Like Ferranti and all the younger generation of electrical engineers, they were, for the most part, 'AC' rather than 'DC' men, and they believed in big power stations serving large areas. Above all, they recognized that if electricity was going to be available cheaply in the home, for transport and in industry, the

British parochial approach to its supply would have to be abandoned. It took nearly half a century for that to happen nationally, but in his native north-east Merz was able to engineer a regional 'grid' on a scale that had proved impossible elsewhere.[3]

Charles Merz was born in 1874 in Gateshead, Newcastle upon Tyne, into a family which was already involved in the infant electricity industry. Though he was just ten years younger than Ferranti, the pace of technological advance was such that he could be considered an electrician of a second generation. His father, the sternly intellectual Theodore Merz, an industrial chemist, was one of the founders in 1889 of the Newcastle upon Tyne Electric Supply Company. His mother, Alice, came from a Quaker family with many business interests in the north-east. The family home, The Quarries in Newcastle, was visited by leading industrialists and bankers. An uncle was a director of the giant shipbuilding firm Swan Hunter. Theodore Merz was a philosopher who produced four volumes of a monumental *History of European Thought in the Nineteenth Century*, a project he was still working on when he died in 1922.

Charles, however, was no intellectual and, like many other young engineers, preferred to get his hands on machinery rather than study the theory of electronics. After a basic education at the Quakers' Bootham School in York, he spent some time at Armstrong College in Newcastle, where his father was a part-time lecturer. However, he did not stay to finish a degree, leaving at the age of eighteen to become an apprentice to the Newcastle upon Tyne Electric Supply Company, working in its first power station. A rival company, the Newcastle and District Electric Lighting Company, had been established by Charles Parsons, also in 1889. Rather than compete, these two companies came to an agreement about territories, Parsons's company supplying the west of the city and Merz's company the east.

As an apprentice, Merz had to be at work at six in the morning and was given many of the routine tasks involved in keeping a power station ticking over. After a while he was sent out to install electricity in homes and offices. After two years he went to work for Robey's of Lincoln, as part of an apprenticeship exchange scheme, where he learned more about engineering. Again it was a six o'clock in the morning start, in Lincoln after a cold bath. Here he met one of the friends who stayed with him for much of his career, T. H. Minshall. The two compensated for their primitive digs with evenings playing billiards at the Great Northern Hotel, their lifestyle subsidized modestly by an allowance from Charles's father.

Theodore Merz had a huge influence on his son. He had had a most unusual childhood himself. His father, Dr Philip Merz, was a German who had come as an educator to Manchester, where he was appointed headmaster of Chorlton High School. Christened 'John Theodore' (he chose to be called by his second name), he was born in Manchester in 1840 but was brought up in Germany after his father moved back there and settled in the small university town of Giessen, where the famous chemist Justus Liebig had established a laboratory. Theodore studied at Giessen before moving on to Göttingen University to continue his interest in mathematics, physics and chemistry.[4]

When he was twenty-eight Theodore came back to Britain to work as a chemist with the famous firm of Charles Tennant in Glasgow, before moving down to Hebburn on Tyneside. One of his brothers-in-law was a prominent Liberal MP, Robert Spence Watson, uncle 'Bob' to young Charles, who was involved with the setting up of the Newcastle upon Tyne Electric Supply Company. Before that, Theodore Merz had been an investor in Joseph Swan's incandescent light company.

After Robey's of Lincoln, Charles Merz moved down to London. Again his father had a hand in his career, for he was a director of British Thomson-Houston (BTH), offspring of the American company. BTH had won a contract to run Bankside power station on the Thames in London (now Tate Modern), and Merz was taken on to supervise the installation of steam engines and other machinery. From there he moved on to Croydon in Surrey to work on the building of a new power station, where he installed some BTH equipment.

As British Thomson-Houston was in business to supply American equipment, Merz found himself working much of the time with American staff. He moved from Croydon to the London office of BTH, where he had a job ordering some of the motors and electrical apparatus for the Central London Railway. As a director of BTH his father was a frequent visitor to London, and the two met up regularly and would discuss his next career move. It came as a shock to Charles – he regarded it almost as an insult – when he was asked to leave London for the town of Cork in southern Ireland, to oversee a new tramway and lighting system to be installed by BTH.

He threatened to leave, to join Siemens, the training ground for so many young electricians. He complained to his father that Cork would be a kind of exile. But Theodore stood firm and persuaded his headstrong son to take the job, as the remoteness of Cork from the technological heartland of Britain (there was no Republic of Ireland in 1898) would be to his advantage. Charles would be in charge, responsible for an entire scheme, though he was only twenty-four years old.

Persuaded by his father to accept the job, Merz took the boat to Dublin and the train down to Cork. At the railway station he hired a jaunting car, a two-wheeled open carriage which was the Irish form of the horse-drawn cab, and told the driver to take him to the 'power house', since he imagined that work on the new power station would be a matter of common knowledge in the town. After a brisk ride the driver pulled up outside the Poor House. It was the first of many charming misunderstandings during Merz's stay in Cork, which turned out to be, as his father had anticipated, a brilliant career move, though not entirely for the reasons he had envisaged.[5]

The Cork project was a little unusual for its time because there were two contracts, one for the creation of an electric tramway and the other for electric lighting. A scheme had been worked out by an

American, H. F. Parshall, who was the consulting engineer on many BTH projects including the Central London Railway. But it was Charles Merz who had to make it work. It was a DC system, with batteries to store power for periods of low demand, and in that sense it was 'old-fashioned'. But Merz oversaw the building of a power station that supplied electricity not only for lighting and trams but also to a small extent for industry, when he persuaded a cold storage firm to switch from gas to electric refrigerators.

The work was far too extensive for him to handle himself and he was soon looking for experienced engineers to take responsibility for different aspects of the Cork project. Once again the firm of Siemens proved to be the chief source of youthful expertise, providing two of the men Merz needed. One of these was William McLellan, who was put in charge of laying the electric mains. He was a few months younger than Merz, born in Scotland in Palnackie, which was a river port for the town of Castle Douglas, Kirkcudbrightshire, before the railway arrived in 1859 and took away the harbour trade. Nothing much is recorded of McLellan's early life, but he was clearly more academic than Merz since he completed a degree at Liverpool University before joining Siemens in 1896. In his 'Reminiscences' Merz wrote: 'I soon took a great fancy to McLellan and this was the beginning of our life's association together.'[6]

It was Merz, in the first instance, who, in 1899, got the new consultancy established while McLellan remained in Cork. An uncle, J. Wigham Richardson, had a shipbuilding business on the Tyne and was also a director of Walker and Wallsend Gas Company. This family connection gave Charles his first big break. The Gas Company wanted to supply electricity, for which it needed parliamentary permission. Charles was asked to present the company's case in London, as he already had considerable experience of electricity supply systems. When he looked at the plans that had been drawn up he was very critical of them. The Gas Company wanted to appoint him as their engineer but he would only agree to act as a consultant. He was far too ambitious to be tied to just one small electricity supply company.

The bid for permission to supply electricity was successful, and Merz began his transformation of the electricity supply of Tyneside and the north-east of England. In Cork he had added to the demands of trams and lighting a small amount of supply to industry. This was now the key to his plans for Tyneside, where both the coal mines and the great shipbuilding yards promised a huge potential demand for electric power if it could be supplied at a price and in a form which was attractive. With his family contacts in the shipyards, Merz preached the superiority of electric motors over steam engines to power many of the specialist shipyard machines. 'There were some prejudices to be overcome; they all wanted to know how much it would cost them compared with the numberless small steam engines they used, though actually it turned out this was not the most important point for them,' he wrote in his 'Reminiscences'. 'The really important point about adopting electricity was that it made available for their operations a much more easily adaptable form of power.'[7] Shipyards that went electric actually spent two or three times more on power then they had previously with steam, but they still cut costs because electric motors could be used for a much greater range of tools.

William McLellan formally joined Merz to form a partnership in 1902. Their ambition, to provide power for lighting, trams and industry, presented them with a huge technical problem. They had to design a power station that could produce electric current compatible with the demands of all three systems. In particular they needed a form of current that could power the induction motors used to drive machinery and could be supplied over a wide area. These motors had been developed in the United States and in both Germany and Italy, but not in Britain. They required a form of alternating current called 'three phase' or 'polyphase', which enabled them to maintain a constant speed under different loads. Nikola Tesla had pioneered the polyphase system in

America at the same time as Galileo Ferraris, working in northern Italy. As with electric systems for transport, the British manufacturers got left behind, and polyphase generators had to be imported when Merz and McLellan built their first power station, Neptune Bank, in Newcastle.

In the history of the electrification of Britain the opening of Neptune Bank was a landmark. It was not that large, but it was the first in the country to produce polyphase AC current, the frequency of which had been carefully judged by Merz and McLellan, after extensive inquiries, so that it could power lighting without flicker, traction engines and induction motors. It was the prototype of all modern coal-fired power stations and was recognized on its opening in 1901 by the distinguished scientist Lord Kelvin as a portent for the future. He said, famously, as the power came on: 'I don't know what electricity is, and cannot define it – I have spent my life on it. I do not know the limit of electricity, but it will go beyond the limit of anything we conceive of today.'

The Newcastle Electric Supply Company (known as NESCo) developed by Merz and McLellan grew rapidly in the north-east. When an application was made to Parliament to introduce electric trams on the north bank of the Tyne, the steam-driven North Eastern Railway objected. Merz persuaded the railway company that it would be better to electrify its lines rather than oppose the electrification of the tramways. This was agreed, and the railway company went so far as to build their own power stations and to rent their excess capacity in power and cables to NESCo. Shortly afterwards, NESCo took over the County of Durham Power Company, which supplied trams over a wide area. In 1904 Merz and McLellan opened a new power station at Carville, at the time the most up-to-date and powerful in Britain, relying exclusively on Parsons's turbines to generate its electricity.

Before the Great War the north-east of England had something resembling a modern system of electricity supply. Small power stations attached to collieries and shipyards were gradually absorbed into a wider system so that the least efficient producers of electricity were closed down.

Though they were not primarily inventors, Merz and McLellan did introduce a number of innovations to power station construction and in the linking of stations so that the load could be distributed efficiently as demand varied. In America and Germany similar interconnected systems were being developed at the same time. Chicago, in particular, had become very advanced, under the management of an English émigré, Samuel Insull, who became a close friend of Merz.

London, however, advanced little at all. The byzantine structure of its local government, through which the laws relating to electricity supply were administered, had prevented any but the most minor and parochial mergers of power companies. In fact there really was no concept of a power company in London: instead, all the major customers for electricity, notably the underground railways, the trams and the lighting companies, had generators for their own exclusive use. The same political decree that had sabotaged Ferranti's brilliant scheme continued to keep the capital and heart of the Empire in the dark ages. It was a state of affairs which engineers like Charles Merz found quite shocking. Once the NESCo system was more or less complete, he and a distinguished consortium got up a scheme to rationalize London's electrical supplies. This meant presenting a Bill to Parliament and, as it turned out, to face not only determined opposition from vested interests, but formidable rivalry as well. Meanwhile, wealthy London was happily installing local power stations that offered the well-to-do the luxury of electric light and, in some cases, electric household appliances too.

CHAPTER 11

ELECTRIFYING LONDON

When electric power was thought of almost exclusively as a way of providing clean smoke-free lighting in the home, it was a luxury and became established in the wealthiest districts of the large cities or in stately homes. At first anyone wanting this new lighting had to find a way of generating electricity themselves, as William Armstrong had done at Cragside. The Grosvenor Gallery station, which grew into the London Electric Supply Company, served the wealthy residents of the West End and fashionable shops. When the law was changed in 1888 to allow private companies forty-two years' grace before they could be compulsorily purchased by a local authority, there was a spate of new, small-scale stations established around the country. Not surprisingly, most of these private enterprise lighting companies found their customers in the wealthiest parts of town. London's West End had a proportionately large number, and in the capital the dominance of private over public ownership of power stations set it apart from the rest of the country.

A typical London power station was the Notting Hill Electric Lighting Company, established in 1891 with the distinguished William Crookes, inventor of the 'radiometer', as a director. In June there was a ceremonial opening of the company's power station, established in the basements of a row of houses bought and converted for the purpose at Bulmer Place.[1] Chairs were set out in one of the rooms occupied by the station's storage batteries, for this was a DC system on the model advocated by Rookes Crompton and Thomas Edison. Crookes, according to the *Daily News* reporter who covered the meeting, was in good form. Not long ago, he told an assembled crowd of shareholders and potential customers, electric light had been a 'dream of his life'. Ten years previously he had installed electric lighting in his Kensington home, and claimed that he was the first in Britain to do so. (He had written a letter to *The Times* in 1882, describing the difficulties of running a petrol engine to drive a generator in his home but arguing that electricity was still cheaper than gas and candles and oil lamps, and would be much cheaper still when he could plug into a power station.)

For Crookes the little power station at Bulmer Place was a dream come true: he would now be able to get rid of his troublesome gas-fired Otto engine and get wired up along with his neighbours. The power station could light 10,000 lamps, and when it had established a second station in the area allotted to it by Parliament it could supply 50,000 lamps. At the outset, however, despite the laying of fourteen miles of copper cable, only 5,000 lamps had been ordered.

'Professor Crookes, who is somewhat of a humorist,' the *Daily News* reported, 'introduced one of the company's most indefatigable engineers... when we saw who he was, we marvelled that a man of the Professor's resource did not whistle, or snap his fingers, or slap his thigh. The engineer was a dog. He was carried in the arms of a fellow electrician. He was deposited on the lecturer's table, from which he gazed right and left at the audience, and which he flapped approvingly with his stump of a tail. Then the ladies embraced him. There are worse things than being a dog engineer. This engineer's business is to carry the "mains" or wires into and through the "culverts" – the small narrow tunnels which are run below the street level.' The *Daily News* recalled a similar cable-laying dog used by Rookes Crompton in Kensington.

Crookes closed his jovial presentation with his favourite theme: electric light might appear to be more expensive than gas light, but when the savings on the destructive quality of gas were taken into account it was cheaper. In a grand home like his own, with valuable paintings, gilding and costly fabrics, the cleanliness of electricity saved a great deal of expense. He had written when he first installed electric light: 'With it the ceilings do not get blackened, the curtains are not soiled with soot and smoke, the decorative paintwork is not destroyed or the gilding tarnished, the bindings of books are not rotted, the air of the room remains cool and fresh and it is not vitiated by the hot fumes from burnt or semi-burnt gas, while fire risk is almost annihilated, as no lucifers are used, and the lamps are high up out of reach.' Electric light also replaced all the candles and oil lamps he had once needed for the rooms where gas was unacceptable.

'Professor Crookes having finished his address,' wrote the *Daily News*, 'Mrs Crookes turned on the current and on a blackboard behind her a pattern of lamps spelled out the initials of the company N.H.E.L.C. 1891.' At a meeting of the company that day, Crookes said he hoped it would exist for the full forty-two years of its parliamentary allotment. That would take it up to 1933. Crookes clearly did not anticipate the imminent obsolescence of his brand new power station in Notting Hill, and appeared unable to appreciate that the lighting of grand homes like his own was just one application of this wonderful new source of power.

In its enthusiastic survey of all the many schemes being introduced in the capital, the *Daily News* considered the position of London's 'Clubland' as a potential goldmine for one of the new lighting companies:

> It is an oblong – bounded on the south by Pall-Mall, on the north by Piccadilly, on the east by St James's-street, on the west by Regent Street... Within this extremely narrow area there is scarcely a house the occupant of which does not belong to the classes from which electric supply companies rely for their customers... Even those splendid clubs alone which render this quarter of London unique in the world might give enough custom to keep an electric company going. The clubs while the rest of the world sleeps, burn the midnight oil – or its equivalent in gas.'[2]

The Grosvenor Gallery station had found some customers in the area, but another enterprise, the St James's and Pall Mall Electric Lighting Company Ltd, had got the Clubland franchise. It would have lit the area five or six years earlier, said the *Daily News*, if it had not been for the 'vexatious restrictions of the Electric Lighting Act', for which Joseph Chamberlain was largely to blame. District by district the wealthier areas of the capital were being wired up by private companies which were busy digging up the roads to lay their mains cables in culverts. The St James's company set its Clubland generating station in Mason's Yard off Duke Street, and put its copper wires in cast iron pipes laid alongside or beneath layers of existing subterranean conduits of various kinds. 'The underground of Clubland,' reported the *Daily News*, 'is more perplexingly riddled and cobwebbed with pipes and burrowed out in arches and vaults than any other quarter of London... water-pipes, gas-pipes, Exchange Telegraph Company's lines, the Fire Brigade lines, the hydraulic mains, the Post-Master General's ditto and other systems threading in every direction...' Once the electric mains were laid, however, and these were nearing completion at the end of 1889, getting wired up was straightforward. 'When a householder or shopkeeper or club secretary in any of these streets wants to give up his gas burners for glow lamps all the company's workmen will have to do is to remove a few inches of pavement, drill a hole through the culvert which passes by the new customer's door and connect the cable inside the culvert with the premises. All this can be done expeditiously and without the slightest inconvenience to traffic.'

One by one the clubs and hotels around St James's Square were getting wired for the new lights. The Marlborough Club in Pall Mall (closed in 1953) was one of the first, and was soon joined by the Continental Hotel in Regent Street. Her Majesty's Theatre in the Haymarket was an early customer of the St James's company and ordered 3,000 'glow lamps', as the incandescent bulbs were often called, for its Christmas pantomime. It was not long before the Café Royal in Regent Street was hooked up to the generator in Mason's Yard. All this was achieved with the DC current favoured by Rookes Crompton, Thomas Edison and many other pioneers of electricity supply. But the St James's company dispensed with storage batteries, confident that it could give a continuous supply without the back-up they would have provided. Theirs were strictly local stations, favourably sited where a concentration of wealth ensured enough custom to make the enterprise profitable.

The *Daily News* took a special interest in the benefits afforded by electric light over gas.[3] Of a new enterprise in west London, a House-to-House venture promoted by Robert Hammond, it reported: 'A good test of the superiority of the new light over the old was afforded by the experience of a fishmonger and poulterer in this region. Light for light, he thought the electric light cost him 20 per cent more than gas. But he found that the difference was more than counterbalanced by the saving in cleaning and repairing and the additional saving in his perishable class of goods. "For one thing," said he, "I can pile up these goods as high as the ceiling." With gas light, and the consequent high temperature in the upper strata of the atmosphere, this would have been impracticable…' Another advantage was that electric lights did not blow out in the wind, and that they were brighter than gas lights and would be cooler in the summer months.

One by one the subtle advantages of electric light over gas were chronicled by the *Daily News*.[4]

> By touching a knob with your finger, you might light up in an instant the recesses of some cupboard or dark corner where the presence of any other light might be troublesome or dangerous. The corners of large drawing rooms may be turned to account by a portable glow lamp, to an extent attainable only after some trouble with any other light. These are perhaps minor considerations; but they are often the very ones which decide the mistress of a house to send for the electrical light fitter. Then there is the worser half – who deserves to be considered sometimes. When in the small hours he lets himself in with his latch-key, it is possible that he may find the passages inconveniently dark. But if they are 'wired' it does not matter… for a touch of a switch may light them up in the twinkling of an eye from basement to top storey.

There was no end to the convenience of electric light. A *Daily News* item noted: 'The electrically lighted looking-glass is becoming a favourite addition to dressing-rooms. The looking-glass opens like a book. Behind the glass portion is a small lamp which can be lighted at any time by the touch of a switch fixed somewhere on the wall. The light cannot come out through the glass… But it emerges through the transparencies which run along the four sides of the glass… Many people have "switches" close by their pillows. This is a luxurious and perhaps lazy use of the new light. During sleepless hours, it must prove handy for the indulgence of reading in bed. And if a burglar were about it might prove less handy for the burglar than for his intended victim.' There was the story of a 'well known lady', the wife of a Lord, who was woken up by a terrific scurrying in her bedroom. 'A touch of the electrical apparatus revealed to her, however, not a burglar, but her favourite terrier in a life and death struggle with a rat.'[5]

At that time, in the early 1890s, many lighting schemes were established in London and around the country. Whether they were owned by local councils or by private investors, there was not much thought of them providing anything other than power for incandescent lamps. One or two companies began to advertise electric appliances that could be bought or hired. They hoped that many of these would be used in the daytime and would therefore 'spread the load' of their power stations over twenty-four hours so that their sales of electricity were not confined so much to the hours of darkness. Rookes Crompton, who established in the 1880s a typical London lighting company at the newly built block of apartments called Kensington Court, was an early manufacturer of such electric gadgets as heaters for hair curling-tongs and ovens. In 1894 the City of London Electric Lighting Company advertised an extensive range of equipment including electric kettles, saucepans, radiators, irons, hotplates and a cigar lighter.

In addition to a new range of gadgetry, the wealthier homes also began to modify and adapt electric light, which many judged far too bright once the novelty of having it installed had worn off. Manuals began to appear on the best way to subdue the glare and lend to electric lighting a certain artistic attraction. In 1891 Mrs J. E. H. Gordon, the wife of a prominent engineer, published *Decorative Electricity*, in which she offered advice on taming the new incandescent lamps:

> Most of the electric light found at present in dining rooms, is very glaring and disagreeable, and fully justifies the remark I so often hear made by ladies, 'I will never have the electric light in my house as it gives me a headache whenever I dine by it' and I am not surprised if they have been accustomed to a light similar to that by which we ate our dinner and tried to converse a short time

> ago. There was a round table seating ten guests, and ten lamps with lemon-yellow shades were hung just above their eyes, so that the light was focussed into the eyes and face of everyone sitting at the table, like a horrid detective little bull's eye, showing up every wrinkle and line in the face.[6]

It was not long before a craft industry was at work, aiming to solve the problem of glaring electricity with delicate silk shades and all manner of decorative ironwork holders and standards. The fact that this was the 1890s and for many years afterwards a luxury trade is nicely illustrated by the advertisements that adorn Mrs Gordon's book. For example, B. Verity & Sons of King Street, Covent Garden, claimed to have 'The largest and most varied selection of high class electric light fittings in London.' And they listed their clientele: 'The Duke of Fife, Earl Rosebery, Earl Cadogan, Lord Randolph Churchill...' and so on through the Knights of the Realm to Coutts' Bank, the Royal College of Physicians and numerous august bodies and West End theatres.

At the same time that electric lighting supplied by essentially local companies established in wealthy districts of London had become commonplace, it was the electrically powered and lighted tram that was the notable feature of the poorer districts of the capital. Whether the trams were run by the London County Council or by private companies, they were largely independent of local lighting concerns. They were not connected, either, to the underground railways. And even in 1914 there were practically no power stations in London supplying power to its thousands of industries, while American visitors were surprised to find the streets still lit with gas lamps.

The patchiness of electric power in London, compared with its widespread adoption in major cities in the United States and Germany, was noted by many visitors from abroad. As late as 1923 the editor of the American magazine *Electrical World*, after a visit to St Paul's Cathedral in London, where he saw an inscription to the 'Unknown God', wrote that electricity was 'almost, but not quite, as unknown' in the capital.[7] New York then generated more electricity than the whole of Britain, and before 1914 the contrast between London and Chicago or Berlin was even more stark. The reason for this backwardness was patently obvious. It had nothing to do with a failure of technology and everything to do with politics. Whereas in most other great cities of the industrializing world a way had been found to amalgamate electricity companies and to supply a range of customers from large central power stations, all attempts to achieve that in London had failed.

Americans, as champions of free enterprise, generally blamed the failure of London, and most of Britain, to modernize its electricity supply on the widespread public ownership of power stations and the dominant political belief that private companies should always be liable to purchase by local councils. While there was a good deal of truth in this assessment of British technological 'lag', it was not quite as simple as that. There is no doubt that the promotion of 'municipal trading' created a large number of local councils fiercely jealous of their right to look after their own 'essential services', such as water, sewage, gas, tramways and electricity supply. In London the creation of the first single large authority, the London County Council (LCC), in 1889, had brought into power a group of political 'progressives' who were keen to be *the* suppliers of electricity to the capital. However, London, to a greater extent than other large cities, had a proportionately large number of privately run power stations like those of the Notting Hill or the City of London companies.

The Board of Trade, in charge of the administration of electricity supply, had in fact encouraged the creation of a multiplicity of stations in the belief that this would bring about competition and bring prices down. Francis Marindin, reviewing the bid by Ferranti to supply a large part of the capital from one power station, encouraged the fragmentation of the industry. There was a ruling that if, in any district, there was a company supplying DC current, then a rival supplying AC could move in to give the consumer greater choice.

At the beginning of the twentieth century in London, therefore, a powerful new authority, the LCC, had ambitions to take over the supply of electricity within its boundaries, or at least to offer to supply local companies that would become distributors, and a very large number of private companies as well as some local councils who had no intention of handing over their businesses to any outside supplier. It was at this critical period in the history of the electrification of Britain that a solution to the problem of localized, and therefore expensive, generation was offered by the consulting engineers who had brought order and efficiency to the north-east of England. Boldly, Charles Merz and his partner William McLellan, with financial and political backing from a distinguished group of men, promoted a Bill in Parliament which, if passed, would enable them to build three huge power stations in London and to offer consumers of electricity power at greatly reduced prices.

In 1901 Merz and McLellan had set up a small London office with the idea that they would get work as consultants to those lobbying Parliament for the right to establish electrical schemes of various kinds. Among the commissions that came in was one to examine the possibility of creating a system that would provide electricity for the whole of the East End of London. Merz took on the job without much enthusiasm, since he regarded it as a poor scheme, though he did his best to get a Bill to create it through Parliament. He failed, but the attempt inspired him to look more closely at London's electricity supply. It was, as everyone agreed, chaotic, but Merz and McLellan had been able to create order out of chaos in the north-east of England and with youthful optimism thought they had a chance to bring the capital's system up to date.

When he worked at Bankside Merz had visited Ferranti's Deptford station and had been very impressed. He and Ferranti had become firm friends, and shared their belief in the need for large power stations supplying large areas and thousands of customers. However, there was another man who influenced Merz a great deal more and encouraged him in his ambition to rationalize London's electricity supply. This was Samuel Insull, fifteen years older than Merz, who had made his name and his fortune in America but always retained his interest in his native country and was regarded by journalists and fellow electricians alike as the authority on large-scale power schemes.[8]

Insull had been born in 1859 in London, where his parents ran a small dairy; they vigorously espoused the Temperance Movement and were members of the non-conformist Congregationalist Church. Insull's father was a rather sad figure, sometime lay preacher, learned but ineffectual, while his mother held the family together; their first three children died in infancy while five survived – Samuel, an elder brother, two younger sisters and a younger brother.

During a brief period of family prosperity when Samuel's father had a salary as secretary of the United Kingdom Alliance, a national temperance organization that acquired government backing, the Insulls moved to Oxford. Here Samuel and his elder brother were educated by students from the university who taught in a private school. This came to an end when Samuel was fourteen, funding for the Alliance ran out, and his father was once again struggling to survive financially. The boys went out to work, Samuel rejecting the suggestions of his father and finding his own work through an advertisement in *The Times* as a clerk in an auctioneering firm in London, a job which, by chance, led him on to a wider world. He was told by another clerk there was freelance work to be had as a shorthand writer or stenographer, and Insull learned quickly, imbued with the self-help ethic of the Victorian Samuel Smiles. He read widely and came across an account of Thomas Alva Edison, just then, in the late 1870s, becoming internationally famous.

Not long after Insull had adopted Edison as his hero and role model, he was fired from the auctioneers to make way for a boy who was the son of one of the owners. Though he was furious about his treatment, his summary dismissal launched him on a dramatic course. He took a job as assistant to a man in the City who turned out to be one of Edison's agents in London. In time Insull met Johnson, the Edison representative who had sent Sprague to the Menlo Park work station. After he had shown himself to be a diligent and knowledgeable assistant working for Johnson, he too was recommended to Edison. In 1881, without much idea of what awaited him, Insull shipped across the Atlantic to become Edison's secretary, administrator and accountant.

Insull had no background at all in electricity, but he learned to run the business and to understand how it might be made more profitable. He moved from Menlo Park to become the vice-president of General Electric, the Edison company that manufactured electrical equipment. From there he moved on to Chicago Edison, then a fairly small, typically DC, lighting company, which he joined as president. From this one small company he built the largest electrical power business in the world within a few years.

The story of how Insull managed to buy up all the small electricity stations in Chicago and persuade manufacturers and the transport companies to buy his electricity generated in large stations is too byzantine to be described here. He certainly took a lead from Charles Yerkes, and was not above bribing and threatening where necessary. It helped that he had a monopoly on the supply of all electrical equipment to Chicago. His ambition was not simply to become rich, though he did amass a spectacular fortune. Insull, like Merz, was an electrical messiah and wanted to show Chicago and the world how it could be most efficiently generated and sold.

He had two very large coal-fired power stations built with the latest equipment to provide his bulk supply. But simply generating electricity wholesale was not the key to Insull's system. He had noted on a visit to the Brighton station in England in 1894 how a metering system not only appeared fair to customers, as it charged for use rather than a flat fee, but gave the power station engineers a clear idea of how the load was distributed. By installing meters that enabled him to monitor the pattern of usage among his customers, Insull devised an ingenious system for distributing power in the most efficient and cost-effective way. The old Edison system of storage batteries was useless for large-scale power schemes. Though Insull began with DC generation, as his system grew he switched to AC. He was able to switch DC to AC with newly inverted converters and to supply power either way according to the requirements of his customers: the streetcars were run on DC, machinery for manufacture required AC. Huge power stations had to be kept working more or less continuously, as they could not be shut down and fired up quickly. It was vital, therefore, that Insull was able to sell as much electricity as possible throughout the day and night. To encourage industries to take his electricity either as new users, or instead of their own power generated on site, he offered cut-rate supplies which served to take up the slack during the day.

At the centre of Insull's Chicago system was a hard-pressed engineer with the title 'load despatcher'. The job, first introduced in 1903, was nerve-racking, as it required continuous monitoring of the output of the stations so that it matched demand, which varied considerably in a largely predictable fashion. The load despatcher learned quickly about the social life of Chicago and how the behaviour of millions of people governed the demand for electricity. At the end of the working day the demands of industry fell away just as the demands of transport rose. This was the prototype for all later integrated power systems, in which supply and demand are matched as closely as possible by constant vigilance and an intimate knowledge of social customs.

There was no technological reason why London and other large cities in Britain should not have had the same integrated power system as Chicago. Charles Merz and his backers did their best to persuade Parliament that there could be huge savings if the capital would only allow the creation of large power stations. But the Bill they promoted in 1905 failed. So did others that offered alternative schemes for centralizing electricity generation. The London County Council put up several of its own proposals to run central power stations which would offer a supply to private customers and local authorities. But these were rejected.

It did not matter to Merz whether the kind of scheme he and McLellan were proposing, and which Samuel Insull urged on London from his electrical citadel in Chicago, was owned and run by the London County Council or by a private company. The technology would be the same, and it was only technology that interested a consulting engineer. In 1907, when Merz was promoting yet another scheme to reform London's electricity supply, he spent some time trying to get the backing of David Lloyd George, who was then President of the Board of Trade. But it was no use: Lloyd George told him, 'My dear young friend, this is not a matter of engineering; it is a matter of politics.'[9]

The parochialism of London politics ensured that neither private power companies, like those promoted by Merz, nor the London County Council could threaten the interests of the capital's existing electricity providers. Inefficiency and higher-priced electricity were preferable to the monopolistic powers of a huge concern like Insull's Commonwealth Edison company. *The Times*, which began to run a column of 'Electrical Notes' in the early 1900s, lamented the backwardness of London and gave regular reports of Insull's views. On 11 January 1911, under the cross-head 'Bulk Supply in London', *The Times* reported: 'Mr Insull is not alone in evincing surprise that in a great centre such as London the old-fashioned method of employing comparatively small units in a large number of stations is still allowed to continue, and it is obvious that the unifaction of the production of electrical energy would result in lower generating costs. That a cheap supply for London would be a great public benefit is a matter of common knowledge, and Mr Insull sees no reason why what has been done in the United States could not be accomplished in Great Britain.'

In 1907 the question of who should be in charge of electric power in London became a lively issue in the county council election. The old school 'Progressives', who thought the county council itself should be the 'bulk' supplier, were successfully opposed by a Municipal Reform faction opposed to local authorities running such businesses. The argument was that the industry was still new and evolving rapidly, any investments would be a gamble and it was wrong to risk ratepayers' money on an enterprise in which the London County Council had no expertise. That expertise was available, of course, from Merz and McLellan, and in 1913 they were approached by the LCC to draw up a new scheme for the council to run. This new effort to rationalize London's electricity supply had not got very far when the outbreak of war with Germany in August 1914 put an end to it.

CHAPTER 12

THE LEGACY OF WAR

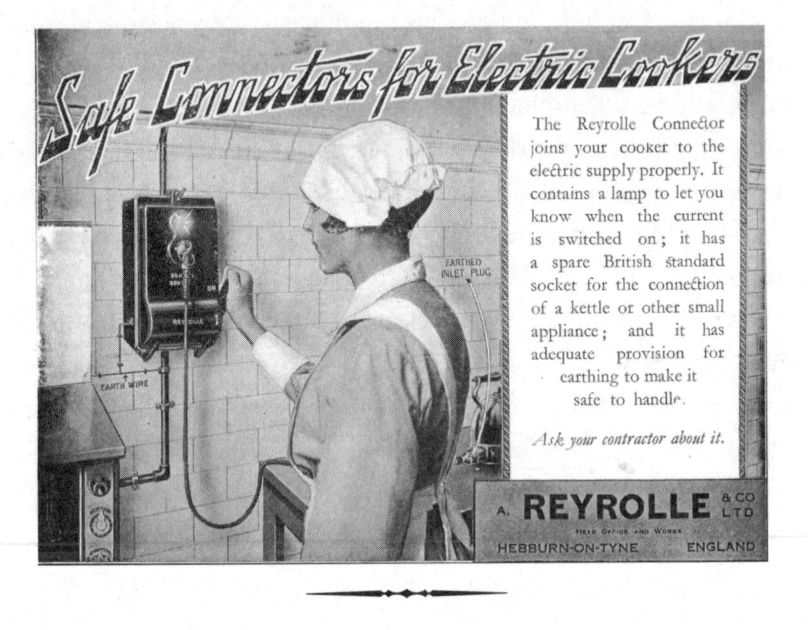

There was no expectation when war with Germany was declared in August 1914 that British industry would be disrupted, nor that there would be any fundamental change in policy relating to the supply of electricity. Notoriously, the nation was optimistic about the prospects of victory, happily repeating the mantra that it would be 'over by Christmas'. Young men, fired with patriotic enthusiasm and a desire to see some action, volunteered in their tens of thousands, leaving coal mines and factories and electricity companies seriously short of staff. By May 1915 it was clear that the war would not be over in a few months and the government would have to take over control of vital industries. A Ministry of Munitions was created, with David Lloyd George in charge.

The war inevitably cut off supplies of electrical equipment from German firms that had competed fiercely with America for British custom. And the German Army was quick to prevent supplies of high-quality steel from Sweden getting to Britain's munitions factories. To make up the loss of imported steel, home production would have to be stepped up and that would mean finding more electric power. Looking around for the men who might be able to get the best out of the nation's parochial power stations, Lloyd George and his civil servants soon lighted upon Charles Merz and William McLellan. Even in 1915 they were still trying to push through legislation to rationalize London's electricity supply, acting as consultants on the most recent scheme proposed by the LCC.

At the outbreak of war the London area, roughly the boundaries of the LCC, had fifty-nine independent distributors of electricity, supplying current in seventy-two different varieties. In 1914 there were thirty-five stations run by councils, after the enthusiasm for 'municipal trading' had taken hold in London as elsewhere. The range of voltages on offer in different districts ranged from 480 down to 110, and the frequencies from 100 down to 25.[1] (Today's standard is 240 volts at 50 cycles per second.) However desirable it would have been to gear up industry for the war effort, there was no possibility of rationalizing London's chaotic supply system.

William McLellan was put in charge of the distribution of electricity and did his best in the circumstances to concentrate supply, and therefore valuable coal stocks, in the most efficient power stations. The major local authority suppliers in industrial areas were encouraged to work flat out and expand their generating power if possible. Companies that ran their own small and inefficient generators were asked to take their power from the larger power stations. The most spectacular response was made by Sheffield Corporation, which supplied the power for steel-making furnaces and increased its output sixfold in the four years of the war. In fact, all the major industrial regions drew huge increases in power from stations which were enlarged wherever possible, and which worked overtime. The final tally was impressive, with 327 municipal electricity suppliers increasing their collective output from 705,000 kilowatts in 1914 to 1,490,000 in 1918, while the 230 commercial stations had nearly doubled their output from 430,000 kilowatts to 788,000 in the same period.[2]

Manufacturers of electrical equipment switched from making meters and transformers to producing armaments. Sebastian Ferranti's firm at first experienced a loss of trade, but quickly reorganized to supply the Ministry of Munitions with shells. In 1896 he had established a works in a place called Hollinwood near Manchester, to manufacture his large generators, and it was this factory that switched to shell-making. By 1918 Ferranti's profits had risen substantially, and he was able with government loans to fit out the works with the most up-to-date American equipment. The firm also recruited a large number of women, who were responsible for an impressive output of standardized shells and fuses.

In their 'Electric Power Supply During the Great War', Arnold Gridley and Arnold Human acknowledged that women had played a significant part in the electrical industry[3] but gave them no credit for their skills. 'From a number of opinions expressed, the employment of women's labour has been found satisfactory on simple repetition process work in mass production, but in other work it has been scarcely encouraging. Boys have shown more initiative and enterprise and were better time-keepers.' In fact the experience of the war led directly to the formation in 1919 of the Women's Engineering Society, from which grew an important pressure group that encouraged the use of electricity and the involvement of women in the industry.

At the same time that the power load to industry was increased dramatically, the smaller lighting companies went through a very unhappy period. Their profits were dependent on the quantity of power they sold, which was in turn dependent on how often their residential and commercial customers kept their lights burning. Early on in the war there was an order to turn off lights at night and to restrict the leisure use of illumination, partly as a kind of blackout measure in the face of raids by Zeppelins, the German airships, and partly simply to save on coal. Then on 21 May 1916 Parliament passed the Summertime Act, which decreed that all clocks should be put forward one hour as a measure of 'daylight saving'. It was an idea that a builder, William Willett, had been canvassing for long before the war and which had been adopted in Germany. In 1916 it was brought in to save fuel and encourage a longer day for industrial and agricultural workers. But for the electricity lighting companies, reliant on the profligacy of their well-to-do customers, it was yet a further blow to their profitability. They had to contend with both a fall in demand and a rise in the cost of coal. In turn this meant that during the war years they had to put up their prices, and electricity became even dearer than it had been in peacetime.

The correspondence columns of the newspapers bristled with complaints from disgruntled customers. 'Will you permit me to draw attention to the action of the Kensington and Brompton Electric Lighting Company in connexion with the Daylight Saving Act?' wrote an irate reader in a letter to *The Times* on 15 June 1916. 'This company, in order to recoup itself for loss arising from lessened consumption of current, is shifting the whole of its burden upon consumers by charging a further 10 per cent upon the price of current... and I maintain that it is not in the interest of public policy that private companies should be able to raise their charges unchecked by Parliament.'

One lighting company after another reported at its annual general meeting a fall in profits and a need to raise charges. Inevitably comparisons were made between the cost of electricity supplied by local councils and that provided by private companies. 'It is notorious,' wrote the anonymous author of the letter quoted above, 'that some London municipalities are providing customers with an efficient service at half the cost of the private companies, and at the same time make a profit which they are able to apply to the reduction of the rates.'

If there were lessons learned from the experience of the war, they hardly took the debate about how best to supply electricity for trams and railways, industries, shops and domestic consumers any further than it had got in 1914. The electrical messiahs McLellan and Merz, now backed up by a chorus of engineering voices, had it right. Electricity was not being used on anything like the scale it could be because it was too expensive. The only way to bring the price down was to generate it more efficiently, which meant in large power stations serving a very large number of customers. And these power stations could be linked together so that as demand rose and fell throughout the day, at different times of year and in different parts of the country, electricity supplies could be quickly switched to where they were needed most. Nobody disputed this, yet political opposition remained solid. Expensive electricity was the price paid for rejecting the potential monopoly power of a few large suppliers, whether they were run by councils or commercially.

Towards the last months of the war, a committee appointed by the Board of Trade, which included both Sir Charles Parsons and Charles Merz, recommended the creation of a new government organization which would take control of the whole industry. Under the headlines 'Cheap Electricity – Proposed Board of Control – Larger Power Stations', *The Times* outlined the radical proposals for its readers: there should be a new body of Electricity Commissioners which would have overall control of the generation of power throughout the United Kingdom.[4] However, existing suppliers, both municipal and commercial, saw no need for change as long as they made a profit to relieve the rates or, in the case of private companies, to satisfy their shareholders. The generating and sale of electricity was not a failing industry, as it had been for a time in the very early days. It was a success story: why interfere with it?

The compelling reason, as far as reformists like Charles Merz were concerned, was that the rest of the industrialized world was rapidly adopting electric power and in many cases was generating it more efficiently. Because it was relatively cheap, it was more widely used than in Britain. In 1919 Parliament very nearly voted to create a new organization with the power to force the myriad of electricity suppliers to reorganize themselves. The country would be divided up into regions for electricity supply, and within these the smallest stations would be closed down and power supplied by fewer large stations linked together, so that one could support the other. The model was Merz's NESCo, which remained the only integrated system in the country. For this legislation to be successful, the hundreds of existing supply companies, both private and municipal, would have to be compelled to co-operate with each other and lose much of their independence. But in the House of Lords, every clause that would have made change compulsory was thrown out. What emerged was an Electricity Commission that had the hopeless task of trying to cajole hundreds of recalcitrant suppliers into abandoning their proud individuality in the cause of the common good.

Though electricity suppliers had made a valiant effort during the war to increase their output, nothing had really been learned. The widespread belief that the demands of the armaments industry had finally forced the country to adopt electric power had a grain of truth, but not much more. It had been impossible to radically reorganize the way in which electricity was generated and supplied. It was not until 1925, when Britain was slipping further and further behind in the electrification league tables, that a newly elected Conservative government led by Stanley Baldwin dared to contemplate the degree of state interference in the affairs of commerce and local government that was needed to modernize the country.

The committee appointed to review 'the National Problem of the Supply of Electricity' was chaired by the down-to-earth Scottish industrialist William Weir, who pulled no punches in his report, published in 1926. He pointed out that the former penal colony of Tasmania had a higher consumption of electricity per head of population than Britain. This was revealed in a startling league table which showed California as the region with the highest consumption of electricity per head (1,200 units) and Britain the lowest (110 units). Chicago was second (1,000 units), followed by Canada, the north-eastern USA, Switzerland, Tasmania, the USA as a whole, Norway, Sweden, Sydney and Shanghai. At this period any country with the potential for hydro-electric power on a considerable scale – and Tasmania is an example – was likely to be ahead of Britain. But, as the Weir Committee pointed out, there really was no reason for Britain to be so far behind other than the absurdly out-dated structure of its power supply: there were 572 authorized undertakings which owned 438 generating stations.[5]

The adoption of most of the recommendations of the Weir Report in a new law passed in 1926, the Electricity (Supply) Act, broke the old deadlock of the parochial control of the industry. It did so by sanctioning a degree of state control that frankly astonished many observers of the British scene. An American Professor of Government, Orren C. Hormell, wrote in 1932:

> The British public is opposed to state operation of utilities, but at the same time, in general, approves of municipalisation of local services... It is not unusual for communities most conservative politically and socially to develop municipal trading most fully. Bournemouth and Blackpool are examples among many where the councils are not only extremely conservative, but also highly successful in municipal trading. Conservative leaders who fight national socialism ' tooth and nail' consider that municipal ownership is not socialism at all – merely 'good business'.[6]

It was a Conservative government, however, that brought an end to the Victorian notion that 'good business' was all that was needed to provide the country with cheap electricity. In 1926 a step was taken towards the inevitable logic of state control.

The Central Electricity Board (CEB) was appointed by government but was to act independently. First it would raise the finance to build a National Grid, linking power stations in nine designated regions of the country. Once this was in place, the CEB would buy all the electricity produced by the nation's power stations (Scotland excluded) and, at a rate to be determined, sell it on to local suppliers, who would distribute power to households, industry and transport systems. Not all existing supply companies, whether private or municipal, would be able to contribute to the Grid. It was the CEB's job to designate those that were most economical in their generation of power. This generally meant those with the larger and more modern power stations. The Weir Committee thought the national total of suppliers would fall from 438 in 1926 to just fifty once the Grid was working.[7]

At the same time the CEB would begin to standardize voltages and frequencies of supply, which varied enormously among the hundreds of suppliers. Unfortunately for Charles Merz, who could justifiably regard the creation of the Grid as a triumph for all his years of campaigning, the Board chose a standard different from that of his company, NESCo, and costly adjustments had to be made.

Nine electricity regions were marked out on the mainland of Britain, running from the lowlands of Scotland down to the south coast. Within each region the engineers of the Board mapped out routes for the cables of the Grid, which would link the main power stations. Britain was at last to be offered cheap electricity for all. But, as with all great technological innovations, there was a price to pay.

PART TWO

BRAVE NEW WORLD 1918–39

CHAPTER 13

THE QUICK PERSPECTIVE OF THE FUTURE

It is doubtful if the British public knew much about the machinations of politicians intent on sweeping away the parochialism of the electricity industry until they watched in awe the construction of giant steel towers as they arose in the countryside, their down-turned arms gripping cables that were strung high above the fields and hedgerows. When they first appeared in the landscape there was still a buzz of excitement about all things Ancient Egyptian, following the discovery of the riches of King Tutankhamun's tomb, which is possibly why these most modern of structures became known as pylons. The name stuck, despite the fact that a pylon is, strictly speaking, a ceremonial gateway, and that electrical engineers continue to refer to them simply as towers.

This was not the first time the country had been carved up by an industrial innovation that stood for 'progress'. There had been canals in the eighteenth century, railways and new roads in the nineteenth century, more roads in the 1920s and a network of poles that carried the telegraph and, later, telephone lines. But the pylons were, by their very nature, uniquely intrusive in any landscape. Those carrying the highest voltages of the Grid were to be seventy feet high. They would dwarf woodland and villages and stride over hill and dale like giants. There was no possibility that the CEB could afford to bury the cables of the Grid except in a few instances. The line taken was determined by engineers intent on efficiency in the transmission of power rather than the impact the pylons might have on the landscape. There was bound to be trouble.[1]

The first approach was made by wayleave officers, recruited mainly from retired military personnel, who would break the news to farmers and villages that the Grid was heading in their direction. There were small compensations: five shillings' rent a year for each pylon, fifteen shillings for a pylon that got in the way of ploughing, five shillings if the pylon interfered with mowing. At every turn the wayleave officers and the Electricity Commissioners emphasized that the whole point of the exercise was to provide everyone with cut-price electricity.

Although some really determined resistance to the 'march of the pylons' was organized in the South Downs, the Lake District, the New Forest and some other parts of the country where the landscape was regarded as especially precious, both by local people and those concerned to preserve what was left of rural Britain, it ended mostly in defeat. In retrospect it is quite surprising how unmoved the politicians and CEB officials were by the protests. The Weir Committee had led to the creation of a state-backed machine that ploughed on in the sure knowledge that however much they were harried about the chosen route for a line of pylons, they could override the objectors. Such was the outcry about the desecration of the South Downs that the government agreed to hold a public inquiry. But it had no appreciable effect. Downland was not sacred, and would have to carry its share of pylons.

In his *Questions of Power: electricity and environment in inter-war Britain*, the historian Bill Luckin puts the failure of resistance down to a number of weaknesses in the protesters' position. First, they were badly organized and some local authorities dithered over whether or not they were for or against the pylons, which some regarded as tantamount to being for or against 'cheap electricity'. Second, there was a genuine difference of opinion between local authorities about the pylons: some were conservationist, others were not. Third, those wanting to steer the pylons away from the South Downs appeared to be happy to see them routed over another area of countryside that might be considered just as precious, in this case the Weald. A case of what would be called in a later period Nimbyism – Not In My Back Yard. Fourth, the phalanx of the great and the good who lined up to object to the pylons perhaps looked rather high-handed, disregarding the opinion of ordinary mortals who might be quite happy to get electricity even if the price was a line of seventy-foot pylons. Last, the protesters were at a huge disadvantage faced with a secretive and determined body like the CEB, which had government backing.[2]

Small battles were won elsewhere: a few cables were buried in the Lake District. The town of Keswick put up such resistance that the CEB simply bypassed it. Ancient laws in the New Forest thwarted the Commissioners over part of their scheme. But overall it went well for the Board. The Grid was put together in an impressively short time. Between 1927 and 1933 a total of 22,000 wayleaves were negotiated or compulsorily purchased (600 of the total) and 4,000 miles of cable hoisted on to the arms of the pylons.[3] The Grid came on stream bit by bit, each region operating independently of the others. Small power stations had to accept defeat: they would no longer be generating their own electricity. But the companies and councils that ran them would still be suppliers, buying from the CEB and selling to customers. It was a sad time for local engineers, who lost their raison d'être. They had not wanted the Grid, and provincial suppliers had opposed the creation of the CEB, arguing that the whole thing was a waste of money.

In the end, however, the choice for the country was pretty stark. Either it accepted the pylons or it remained backward, a nation only fitfully electrified as it was in the 1920s. And not every aesthete regarded the pylons as a blot on the landscape. Considerable thought went into their design, and a myth arose that the CEB had employed an 'eminent architect' to come up with the most pleasing structure. Piqued by repeated reference to himself as the creator of the pylon, Reginald Blomfield had to write to the newspapers on more than one occasion to put the record straight. In responding to a letter in *The Times* objecting to pylons being put up on the South Downs, he wrote:

> The masts were not designed by any architect, 'eminent' or otherwise and the plain facts are these: – The masts were designed by the very competent engineers of the Central Electricity Board, and when the designs were completed the Board consulted me as to their general form and colour. For colour I advised green; in regard to form I suggested an alteration in outline which was impracticable on account of the costs, but another modification proposed by me was adopted by the Board... All who know Sussex and are interested in it, as I am myself, would sincerely regret any injury to the scenery of that beautiful country, unless there are compelling practical reasons, but I would suggest in the first place that such reasons may exist, and that this is a point which can only be decided by expert opinion and secondly that the overhead method of transmission has been generally adopted all over the Continent, and anyone who has seen these strange great masts and lines striding across the country, ignoring all obstacles in their strenuous march, can realise without a great effort of imagination that these masts have an element of romance of their own. The wise man does not tilt at windmills – one may not like it, but the world moves on.[4]

The eccentric artist Eric Gill wrote in support of Blomfield:

> I write not only as an artist but as a Sussex man – born and bred – to whom love of the South Downs is as natural as it is enthusiastic. Anyone who has seen the aqueducts striding, almost galloping, across the Roman Campagna must have been struck by the inexorable majesty of them, and the need of Rome for water is analogous to the modern world's need for power. In France I have seen these great electric standards striding across the country – delayed by nothing. In England, in Buckinghamshire, on a small scale the same thing may be seen. Are we to suppose that beauty is only to be found in certain recognised 'styles' of architecture? Is the Forth Bridge ugly because it is not built of stone? Is the Tower Bridge beautiful because the citizens of London, remembering the proximity of the Tower, saw fit to clothe its iron work in machine-made Gothic? Such an attachment to 'Nature' which goes with a refusal to see beauty in engineering, while making use of engineering and making money by it, is fundamentally sentimental and romantic and hypocritical. Let the modern world abandon such attachment, or let it abandon its use of electric power.[5]

In 1919 the responsibility for electricity had been given to the Ministry of Transport, an odd arrangement but not entirely illogical, as it was thought at the time that a major customer for the power stations would be the mainline railways. Steam survived a great deal longer than anticipated, but electricity remained with Transport, and the Minister when the Grid was going up was Herbert Morrison. Like Blomfield and Gill, he took a robust view of pylons on the South Downs: 'They have a sense of majesty of their own and the cables stretching between them over the countryside give one a sense of power, in the service of the people, marching over many many miles of country.'[6] As the Grid was nearing completion, the pylons entered the poetic consciousness of the young Stephen Spender, who in 1933 attempted to capture their startling appearance in a few verses. Strangely, in the second stanza he had them constructed of concrete rather than steel, but his carefully selected words were telling nonetheless. The pylons are 'bare like nude giant girls that have no secret' and '...like whips of anger, With lightning's danger, There runs the quick perspective of the future'.[7]

In his *Questions of Power*, Luckin characterizes those in favour of the Grid and the greater use of electricity as 'triumphalists', in contrast to the 'traditionalists' who wanted to protect the landscape from the pylons. Whether or not the term fairly characterizes the likes of Gill and Blomfield, the completion of the first stage of the Grid, as planned in 1926, was treated as a great success story rather than an environmental disaster. A *Times* correspondent, reporting from Bournemouth on 5 September 1933, described the scene as the last pylon was put up: 'The raising of the last tower of the grid system of the Central Electricity Board at Breamore, near Fordingbridge, on the out-skirts of the New Forest, today was remarkable for the display of rapid and efficient workmanship. Representatives of the Board and 10 skilful workmen erected the steel tower, which is 70ft high and weighs four tons, in an hour and a quarter.'[8]

The pylon rose so rapidly because it was in kit form, with whole sections already bolted together on the ground, to be raised one atop the other with a derrick and pulley system. Anticipating this day, *The Times* of 5 September had greeted the completion of the project with resounding approval: 'When the electricians climb down from a 70ft high steel tower... at 11 o'clock this morning the greatest scheme of its kind in the world – the grid scheme – will have been completed. The tower is the last of the 26,265 pylons that have been built by the Central Electricity Board over Great Britain... The five-and-a-half year scheme has cost £27,000,000 and has meant employment, directly or indirectly, for 200,000 workers.'[9] It was no doubt to the great advantage of Luckin's 'triumphalists' that the grid was put up during the years of the Depression, when unemployment reached alarming levels in the traditional industrial regions of the country.[10]

The Times reeled off the statistics: 4,000 miles of transmission lines, of which nearly 3,000 carried a massive 132,000 volts, the rest 66,000 or 33,000 volts. Most towers were around seventy feet in height but some were monumental: those carrying cables across the Thames at Dagenham were 487 feet high, the largest of their kind in the world at that time. By 1933 just over half of the Grid network was delivering electricity and it was hoped that by 1935 it would be near to 70 per cent complete. At a kind of 'topping out' ceremony when the last pylon had gone up, Mr J. W. Beauchamp, CEB manager for the south-west of England and South Wales, made a short speech. He mentioned, in passing, the 'little trouble' in regard to the New Forest, but said most landowners and tenants had been helpful. In any case there was 'so much natural beauty in England that they would have to put up a lot of grid lines before they began to spoil it'.[11]

Beauchamp had a vision. He believed, *The Times* reported, 'the next 10 or 20 years would change entirely the nature of home life in this country, the outlook of the housewife, and the health of the people by the bringing of electricity. The grid was only the backbone of the great structure. They were looking most earnestly to the 500 or 600 corporations, councils, and trading companies in the country who had electricity supply powers to follow up the grid, and, casting their bread upon the waters, run out thousands of miles of lines to distribute electricity.'[12]

Listening to this speech, and glancing up at the pylons marching across the countryside, it would have been reasonable for the locals in the New Forest to imagine that all this effort and expense had been to bring electricity to villages and farms. But this was not the case at all. The pylons simply linked districts with existing high demand to where the most efficient power stations were sited. The connecting cables hung over isolated farmsteads whose owners might get their fifteen shillings a year as compensation for each pylon that interfered with ploughing, but no supply of electricity. In 1939 only 50,000 farms in Britain had electricity, about 12 per cent of the total. Most of the larger villages with populations of 500 or more had a supply, but there were still districts in the countryside and in the poorer parts of the major cities that were without.[13]

In contrast to the slow adoption of electricity in the home, public displays of electric lighting became an exciting feature of town centres in the 1930s. In the early days of electricity supply the expectation had been that it would replace gas for street lighting, as it did, temporarily, in a number of places. However, carbon arc lamps were expensive, and the little carbon filament lamps were not really suited to lighting large spaces. Gas returned to the streets and dominated street lighting well into the twentieth century in most towns, while the filament lamp gradually ousted the gas mantle from the home. Incandescent light bulbs were greatly improved after the American Irving Langmuir found a way to make a fine wire filament from tungsten, and these new lamps were marketed from 1914. They gave a very bright white light and began to appear on the streets of London and other cities.

A vogue for 'floodlighting' arose in the 1930s, and huge crowds would gather to wonder at the illuminations. On 1 September 1931 *The Times* reported: 'A rehearsal last night of the floodlighting of London indicated the brilliant effects that may be expected to-night when the full programme will be put into operation for the opening of the International Illumination Congress. Electric floodlights, each of 3,000 candle-power, will throw their beams on Buckingham Palace, and amber-coloured light will give brilliance to the Victoria Memorial.' Not all the illumination was with electricity: St James's Park was lit with gas lamps, but this was a turning point in the lighting of streets and public places. The following year a new lighting system was installed in Piccadilly Circus, London's great entertainment 'show room'. Arc lights had lit the Circus early in the twentieth century, but in 1910 gas lighting returned. In 1932 gas gave way once again to electricity.

Though the Central Electricity Board controlled the bulk supply of electricity, it could not dictate who it was sold to by local suppliers. It was up to them to offer connections to householders and companies in their immediate district. The assumption of the Weir Committee and the Ministry of Transport was that a host of mini-grids would extend like a web around the local supplier, which might be a local authority or a privately run company. To some extent this did happen: the miles of low-voltage cable rose nearly sixfold between 1929 and 1939, from 3,700 to 20,000 miles. But these local supply companies really had no incentive to seek out new customers if it meant heavy expenditure on cables. Isolated farms or Victorian housing stock occupied by poor families in the towns were not attractive propositions. For the local authorities it might mean raising the tariff paid by existing customers, or raising the rates, to fund the new connections. For private companies it was simply unprofitable.

Some municipal undertakings, for example Manchester, introduced pre-payment meters for the less well-off customers, while demanding that a condition of supply was that they used at least 100 kw hours a year. This was the other problem with getting a supply to householders in poorer areas: they would want electricity only for lighting, and that did not provide a sufficient return on investment in cabling. It was no use, either, asking these families to pay the £5 or £6 it would cost to bring wiring into their homes from nearby mains. However, some success was had with 'assisted wiring schemes', in which the cost of installation was spread over a number of years so that there was no capital outlay. At the same time both municipal and private suppliers were allowed by law to rent electrical appliances to poorer families. But even in the 1930s, when the Grid was beginning to bring down the cost of electricity, the anticipated revolution in domestic life that this was supposed to bring about had barely begun for the majority of the British people.[14]

As the propagandist Eileen Murphy put it in her 1934 pamphlet *The*

Lure of the Grid: 'Great central generating stations have been built and equipped with expensive machinery; steel pylons have been erected all over the country to carry electrical energy to houses, factories and farms. Only one thing is lacking. There are not enough people who are willing to buy the electricity that is produced...'[15] The reason, according to Murphy, was straightforward: electricity was quite unnecessary and it was, in any case, unhealthy, as there were no draughts created as there were with coal or gas fires. Electric cookers were no good: they were slow, unreliable and the boiling plates were fitted with only three temperatures. And as for hot water: what nonsense was claimed about getting it constantly and instantly. The only proper use for electricity was in factories and decorative neon lighting. It had no place in the modern home.

The Lure of the Grid was just one of many broadsides issued between the world wars in the battle for the hearts and minds of the British housewife, who, by a strange twist of fate, found herself a central figure in the debate about the future energy policy of the country. Each year the manufacturers of electrical equipment were bringing out new products billed as 'labour-saving' devices designed to appeal to women, who, it was assumed, shouldered the burden of all domestic chores. The gas industry, with its established range of cookers, water heaters and fires, proved to be a tough and resilient adversary. It was familiar, the costs were well understood and it was regarded as 'traditional'. Electricity was daringly modern and 'scientific' and futuristic. Though it was regarded as expensive compared with gas, the true cost of using an electric cooker or water heater was not widely known.

A group of manufacturers and supply companies had formed a kind of public relations company, the Electrical Development Association, in 1919 to encourage the use of electricity, but it was a feeble organization with very little funding. Rather more effective, perhaps, was the Electrical Association for Women (EAW), which recognized that the most influential person as far as domestic supplies were concerned was the housewife. In retrospect this Association appears quite bizarre, one of the strangest pressure groups in British social and economic history. But it was very much of its time.

CHAPTER 14

WIRED WOMEN AND THE ALL-ELECTRIC HOME

In her foreword to the first brief history of the EAW, of which she was then president, Lady Moir wrote: 'The Electrical Association for Women might be likened to Electricity itself, in the speed and ease and brilliancy with which it has illuminated the lives of other women in the short space of ten years... It has been dispersing some of the darkness of the drudgery for the housewife; it has been giving the school teacher electrical knowledge; it has been helping the authorities realize the woman's point of view regarding the great scientific development which is affecting their social lives.'[1]

The EAW was formed in 1924 as an offshoot of the Women's Engineering Society, which was the inspiration of Katherine Parsons, the wife of Charles, inventor of the steam turbine. It was the experience of women working in munitions factories during the 1914–18 war that had brought the realization that they were quite capable of handling machinery and understanding how it worked. There was a determination that in future they should not be excluded from what had been an entirely male preserve.[2] Lady Parsons, as she was titled after her husband's knighthood, employed as the first secretary of the Women's Engineering Society in 1919 Caroline Haslett, then twenty-four years old, with some experience of engineering behind her.

Born in Worth, Sussex, in 1895, Haslett was the daughter of a founder of the co-operative movement, Robert Haslett, who, with his wife, Caroline, had five children. Robert was a railway signal fitter, his daughter Caroline a clever and enterprising woman who was to move in socially elevated circles as she pursued her interest in the promotion of a feminine approach to electricity. When she left Haywards Heath high school in 1913, Haslett took a clerical job with the Cochran Boiler Company. Bored with paperwork, she persuaded the firm to allow her to train as an engineer. Electrical manufacturing held out many possibilities for women. In factories they mostly took on work that required dexterity but was not physically demanding. In the home electrical gadgets offered a prospect of taking some of the effort out of the painstaking work of cleaning and washing. Haslett not only wanted women to grasp what she saw as the advantages of the new electrical world, she wanted them to understand it. She wrote a number of 'Teach Yourself' books that explained how motors and generators worked.[3]

In 1924 a group from the Women's Engineering Society formed the Electrical Association for Women and made Haslett the first director, a post she held for an astonishing thirty-two years. She had to resign in 1956 as her health was failing. It is not easy in retrospect to characterize Haslett's organization: was it feminist, as was often suggested, or really retrogressive, since it tended to accept the position of women in society as the home-maker? Whatever else it set out to be, the EAW became, perhaps inevitably, quite comically patronizing in its efforts to persuade 'ordinary' women that they would be better off with an all-electric home. Although it was proud of its few working-class members, most of those involved in demonstrating the use of electric cookers and lecturing on the scientific management of the home were middle-class or aristocratic.

Something of the flavour of the EAW can be found in the chapter on 'Women with Electrical Homes' in Peggy Scott's little history of the organization, *An Electrical Adventure*. There was the Dowager Lady Swaythling, who had not a 'cold corner in her house' since she had got electric radiators, while Lady Mount Temple was especially pleased with the 'beauty of the light in the right place', and Mrs Herbert Morrison, wife of the Minister of Transport in charge of government electrical policy, found that electricity gave her 'time for public work'.

It appears that the possibilities presented by electricity had quite turned Lady Swaythling's head: her old coal range had been replaced by an electric cooker she called the 'Dreadnought', with two grills wide enough to 'grill a turkey' and a hot-cupboard that could keep warm a banquet for fifty people. Buried in the centre of her table was a cable so that she could have a light enclosed in antique glass that sparkled 'like sunlight on a lake' and brightened the gloom of the Jacobean panels.

Electricity was especially helpful to the owners of large country homes, most of which were too remote to be connected to a town gasworks. For heating and cooking they still relied on coal or wood, and for lighting on oil lamps and candles. Both Sebastian Ferranti and his wife Gertrude were enthusiastic supporters of the EAW and promoters of the ideal of the ' All-Electric Home'. In 1913 Ferranti had made enough of a fortune from his electrical equipment business to buy a minor country house, Baslow Hall, near Chatsworth in Derbyshire. Though it looked a couple of hundred years old, the Hall had been built in 1907 but without any modern conveniences. During the 1914–18 war, the shortage of servant girls meant that Gertrude and her daughters had to take on a great deal of the domestic chores, carrying coal and lighting fires. It was a hard time for this enterprising family: their eldest son Basil, who had won the MC, had been wounded in the trenches and died in France before they could visit him. The rural idyll of Baslow Hall was a sad place in 1919 when Sebastian – always referred to by his wife as Basti – began to turn his mind to relieving the toil of housework with a variety of electrical installations.

The local Notts and Derbyshire Electricity Company did not supply Baslow Hall, so Ferranti had to install his own generators. As Gertrude recalled in *The Life and Letters of Sebastian Ziani di Ferranti*: 'Nobody but myself and members of the family living at Baslow Hall during those years can realise how these improvements temporarily disorganised the domestic routine. The continual alterations made it necessary to plan the cooking a day ahead. Frequently the current would have to be cut off unexpectedly, and we would have to sit in our fur coats for two or three hours while Basti patiently worked away putting things right.'[4]

When the teething troubles were over, the long-suffering household began to enjoy a degree of electrical modernity that was unique for the time. Gertrude was in charge not just of a sizeable country house, but of a farm as well, with a dairy herd and chickens and hay fields. Basti, with fanatical zeal, set out to electrify just about everything. He was far more ambitious than William Siemens had been when he electrified his home, Sherwood, as he was able to experiment with a much wider range of equipment that had come on the market.[5] One novelty, which seemed like a good idea at the time, was to 'store' energy in the form of hot water: a 1,000-gallon tank was installed in the cellar. 'Basti was always convinced that, to ensure cheap electricity, it must be stored in some form and the only way, at that time... was in the form of hot water,' wrote Gertrude in her biography of Ferranti. 'The day, he thought, would come when every house would have a large boiler that could be heated by electricity at night.'

Ferranti's aim was to have his generators working at greatest efficiency at all hours of the day and night. He had thermostatically controlled pumps and valves for a central heating system, designed to pick up the load when it dropped off from other uses. In time just about everything in the house was electric: cookers, washing machines, clothes driers, bed warmers, vacuum cleaners, floor polishers and an 'electric clothes brush'. Ferranti was very interested in wireless and had it fitted in every room. On the farm the dairy had an electrically driven cutter for chopping up feed, a churn for the butter, a 'hay drier' and – disastrously at first – an American-made 'electric brooder' which, according to Gertrude, 'electrocuted a number of chickens'. At night the tennis court was lit up electrically, and Ferranti experimented with an electric lawnmower.

By the mid-1920s Baslow Hall was using so much electricity that it was a relief to the Ferrantis when they were connected to the local supply company. A few years later things were further improved when the Hall was wired up to the National Grid, for which Ferranti had provided much equipment. That was in 1929. A year later, Sebastian fell ill and died while on holiday with his family in Switzerland. Gertrude became an enthusiastic promoter of electricity in the home and one of a number of wives of leading engineers who sought to persuade others that the all-electric home was the thing of the future. In the October edition of the *Electrical Review* of 1928, she had published a piece that very nicely defined the view of electricity of the well-to-do housewife in the twenties and thirties. Gertrude explained that she had not realized what hard work servants were expected to do until the war years, when the girls disappeared, taking 'the pleasanter jobs in the factories'. Faced with a daily drudge with coal, coke, lamps and candles, she realized 'how ridiculously hard and dirty work we expected our workers to do'.

'We have now been using electricity on everything in the home for four years. The difference is enormous! Where we used to require five people to work for us, we now require three, and whereas the cost of washing done at a laundry was a heavy expense it is now nearly all done at home without extra labour.' Electricity made the work of servants easier – they could get up later because there were no fires to be laid in the morning and no coal to be lugged about. 'I have been surprised to see how interested the maids are in their electrical appliances,' wrote Gertrude. 'Naturally, one has to teach them to use the various things correctly, and how to keep them clean and oiled.'

Echoing, no doubt, the view of her husband, Gertrude felt the electricity companies were not doing enough to make it easy for housewives to adopt electric appliances. There should be more schemes to hire out equipment: the gas companies were much better at this and much better at 'attending to the wants of their consumers'. In particular there were all those women whose lives could be made so much easier if only they were taught about electric appliances. 'Think of the married women with families living in small houses, who get little or no help in their homes, and not only *have* the children, but often do all the work of the house themselves. How they do it using coal... it is nothing less than a misery; no man would do it.' In conclusion, Gertrude Ferranti commended the work of the Electrical Association for Women (EAW), which her husband had supported since its formation in 1924.

Margaret Morrison, wife of Herbert, who had been responsible for electricity as Minister of Transport from 1929 to 1931, was a more modest user of electricity. Scott wrote: 'The whole of the housework of her home at Eltham is speedily completed with the aid of an electric vacuum cleaner, duster and polisher. Three outlets have been placed in each sitting-room and two in each bedroom, so that the cleaning appliances can be conveniently used, as well as iron or lamp, fire or wireless. Mrs Morrison does her own cooking, and she finds that she can cook a three-course dinner in one hour on the electric cooker.'[6] Sadly, Herbert was not around much to enjoy this modern domestic bliss as, according to his biographers, he saw little of his wife and daughter in this period of his life.

The EAW was well aware that, even in the mid-1930s, with the Grid beginning to run effectively and the bulk cost of electricity falling, it was not sought after everywhere, and where houses were wired up it was for lighting only. In an effort to discover why there was so little take-up, the Association conducted a survey of electricity in working-class homes, which it published in 1935.[7] At a rough guess, based on the number of smaller houses wired for electricity in 1934, they thought there were probably still 5 million working-class homes without any connection at all. By that time the spread of semi-detached suburbia, especially in the south-east of England, had led to the automatic electrification of a great many middle-class homes, as the new houses were built with plugs and sockets. It was the older housing stock that was left behind.

The EAW asked: 'Is Electricity Popular in Working Class Homes?' It did not have much to go on. The *Architects' Journal* had a supplement in which tenants of new model flats built in the poor district of St Pancras in London had been asked what they thought of electricity. All the flats were wired and the local authority that supplied electricity had a scheme for hiring various bits of equipment. Of the 228 tenants, only one said they would have preferred gas for lighting. A larger number, twenty-five, preferred gas for cooking, while thirteen did not care either way. For heating 203 preferred electricity to gas. The EAW concluded that the working-class prejudice against electricity was a myth. It was just a question of price. The next question their report attempted to answer – again with rather skimpy information – was what did working-class women want most from electricity?

They looked at the hire of electrical appliances in six different London boroughs that had such schemes: Bermondsey, Fulham, St Pancras, Stepney, West Ham and Westminster. It was assumed that anyone hiring equipment rather than buying it was likely to be working-class. By far and away the most popular items in Bermondsey and Fulham were the electric irons. The very first electric irons were cumbersome – Rookes Crompton was selling them as early as 1895 – but by the 1930s there were quite efficient models such as the American Hotpoint. The iron could be plugged into a light fitting, so there was no need for an extra power point in the house.

On hire in one or other of the six London boroughs were electric cookers, kettles, fires, water heaters and washing machines. The take-up varied considerably – very few water heaters or 'wash boilers' were taken – but it was still just a small minority of potential customers who were prepared to pay the extra cost. In Bermondsey, for example, only about 5,000 appliances were leased to more than 13,000 households that already had electric lighting. Whichever way the EAW study tried to characterize the extent of ownership of electrical appliances, the answer was much the same: the take-up was disappointing. Vacuum cleaners, for example, were hardly used, though by the 1930s they were a common feature of most American households.

The EAW had demonstrators who toured the country showing women how cookers and washing machines worked, and it was this contingent of its members who provided some account of what working housewives would most value if they 'went electric'. Two typical potential customers were described by the Association – Mrs A, a countrywoman relying on coal for all her cooking and heating, and Mrs B, a working-class woman in town who had gas for cooking and heating and a coal fire to heat her water. If each were offered an electricity supply, what would they want first?

The imagined order of priorities for the countrywoman, Mrs A, was: lighting, iron, vacuum cleaner, cooker, pump, water heater, kettle, washing machine, wash boiler and lastly radiator. For townswoman Mrs B it was: lighting, iron, vacuum cleaner, cooker, washing machine, kettle, water heater, radiator and lastly wash boiler. All this was, of course, pie in the sky. Even if they could afford the hire of one of these appliances, the cost of running them was likely to be prohibitive. Families who hired radiators, for example, tended to use them very sparingly. In fact in the 1930s only the well-to-do used electricity in a profligate way: most families were very careful to keep their bills down, so that the domestic demand which the Grid was there to service was much more sluggish than the promoters of 'cheap electricity' would have wished.

The EAW put a lot of blame for the low take-up on the manufacturers of electrical appliances. Adrian Forty points out in his book *Objects of Desire* that the gas industry was well established by the 1920s and had developed a much wider range of domestic appliances than the electricity industry. There was no doubt it was cheaper than electricity even in the 1930s, and it was simply better for heating water – the sudden roar of the Ascot heater fired up to run a bath or to fill the sink for washing up was a familiar sound in the majority of homes. Even if electricity had been the same price as gas, what Forty calls the 'millenarianism' of the industry, with its slogans 'To Electricity belongs the Present and Future' or 'Science's Greatest Gift to the World – Electricity' were easily refuted. Most electrical appliances were badly designed and just plain ugly, the cookers especially.[8]

American manufacturers such as Hotpoint and Hoover were catering for a mass market long before their British rivals, and led the way with relatively cheap, well-designed equipment.[9] When a tariff barrier held back imports, the Americans built their factories in Britain, creating a dazzling industrial zone to the west of London which gave J. B. Priestley, leaving on the newly built Great West Road in the 1930s, the impression that he was rolling through California. Just to the north on Western Avenue was the headquarters of Hoover, whose salesmen set out to knock on the doors of semi-detached London. Most of the new housing was wired for electricity, though not necessarily with power points. In older housing connection was not guaranteed, a treacherous situation for the vacuum cleaner salesman in the mid-1930s. Sam Tobin recalled for a television series in the 1980s his experience selling Hoovers in north London at that time: 'You knocked on the door. "Good morning madam, I represent the North Metropolitan Electric Power Supply Company" – that was the authority in the area prior to nationalization. "I have been sent along because you are entitled to have one of your carpets and some of your furniture cleaned with the latest Hoover vacuum machine." Very often it would be "not today thank you", but if you got inside you would follow a script almost down to the last letter that they taught you in a training course. I laid out the Kapok, I laid out the sand in strips, and I laid out the saltpetre, plugged in the Hoover and cleaned them up, trying to charm and persuade the housewife all the while.' There was an occasion when Tobin, having set out his 'three kinds of dirt', asked where he could plug the Hoover in. To which the woman replied: 'Nowhere – it's gas!'

Locally recruited salesmen were taught American-style hard selling. Each morning before setting out for surburbia they sang the Hoover song, to the tune of 'The Caissons Go Rolling Along':

> All the dirt, all the grit,
> Hoover gets it every bit,
> For it beats as it sweeps as it cleans.
> It deserves all its fame, for it backs up every claim,
> For it beats as it sweeps as it cleans.
>
> Oh it's hi-hi-hee, the kinds of dirt are three,
> We tell the world just what it means,
> Bing bing bing, Spring or Fall,
> The Hoover gets 'em all,
> For it beats as it sweeps as it cleans.

'It could be pretty soul-destroying though,' Tobin recalled, 'because you could go for weeks without a sale, and if it was bad weather or if Electrolux or newspaper subscription salesmen had done your territory, it was very difficult to get a demonstration anywhere.'[10]

By the late 1930s even the EAW was beginning to accept that the 'all-electric house' was to remain a dream rather than a reality. They had had one purpose-built in a suburb of Bristol as a kind of laboratory to test various pieces of equipment and as a utopian showcase. In addition to the familiar appliances for cooking and washing, there were electric cigarette lighters, a toaster, a 'small food beater', a milk warmer, a dusting machine, a hot plate, a coffee percolator and something described as a Hotlock trolley. There was concealed lighting everywhere, even in cupboards, phones in the bedroom and portable fires that could be used to warm up chilly corners. No chimneys were needed, of course, and the whole place was spotless. The cost of running it, however, was quite evidently beyond the means of the majority of households at the time.

Nevertheless, in the years leading up to the outbreak of the Second World War in 1939, the consumption of electricity rose steeply. The Weir Committee had set a target of 500 kilowatt hours per head once the Grid was operating, and this was reached in good time. Most of the increase in consumption, however, was in middle-class homes, especially those of the newly built semi-detached suburbia, which had electricity laid on from the start. Only the electric light and the electric iron became familiar across a wide social spectrum.

The reach of electricity supply increased rapidly in the 1930s, but Britain had a great deal of catching up to do. It was estimated that in 1929 only 20 per cent of homes were connected, compared with 45 per cent in Germany and 60 per cent in the United States. By 1933 another estimate had 70 per cent connected in both Germany and the US, 60 per cent in France and Italy, but only 50 per cent in Britain. A further reckoning in 1935 had France (95 per cent) and Germany (90 per cent) still way ahead of Britain at 70 per cent. The continuing parochialism of the supply industry despite the creation of the Grid was part of the cause of Britain's tardiness, but so too was the competition from an efficient and well-organized gas industry which had retreated on lighting but was outgunning electricity with cooking and heating.[11] It was estimated that for every electric cooker in Britain in 1939 there were eight or nine gas cookers.

There would be no all-electric future. But there were already indications at the outbreak of war that electric power had become an integral part of the British way of life, not only for the well-to-do but in a small way for much of the nation. Once the Grid began to work in the regions, those in the control rooms balancing the power output with the demand began to discern patterns which were at first puzzling. It had been known from the very early days that the peak load was in the early evening, when people got in from work, and that it naturally reflected the hours of darkness before people went to bed. A really dull day might also bring about a surge in demand for lighting. But there were subtler variations.

The control room for the south-eastern grid was established first alongside Bankside power station (now Tate Modern) in a makeshift building known as the hut. In the 1930s the control room noticed that there was a surge in demand on Mondays, sometimes around five o'clock in the afternoon but at other times much later in the evenings. It was puzzling and difficult to predict. They always watched the weather – as the Grid controllers do today – and noticed that the 5 p.m. surge occurred on fine days and the later surge on wet or overcast days. After a time they realized what was happening. Monday was washday. If it was fine, the clothes on the line would be dry and ready for ironing in the early evening, producing a surge in demand on top of the normal upward curve. However, in damp weather the clothes would take much longer to dry indoors and the ironing would happen later, when demand generally was falling away.

As more and more radios were run off the mains rather than batteries, popular events began to show up on the Grid and the controllers had to be wary of sudden surges at otherwise quiet times. When the fight between the Welshman Tommy Farr and the black American boxer Joe Louis was broadcast in the early hours of the morning in August 1937, each of the seven Grid controllers had to cope with a massive surge as lights and wirelesses went on around the country. At the time the regions were not interconnected, and if one had failed the other would not have been able to take up the load. There had been such a failure in July 1934, when a series of accidents left London and a large part of southern England without power on a Sunday morning. The *Daily Mirror* ran a headline: 'Grid Failure spoils Sunday Dinners... Electrical Power Breakdown leaves Housewives unable to cook... Trains and trams stopped.' The loss of power had also silenced church organs and inconvenienced a large part of the country in a way that would have been quite impossible before the Grid was constructed.

By 1939 the electrification of the country was advancing steadily, if not quite as quickly or comprehensively as the protagonists would have liked. The Grid had brought the price down and enabled electricity to compete more effectively with its great rival gas. Though it was still operating on a regional basis, an unofficial linking up of all the systems worked smoothly, and though the engineers involved in this unauthorized experiment were reprimanded, it boded well for the future. A National Grid, in theory, would be much less prone to power cuts, as one region could provide back-up for another. To power this supergrid, which was heralded as the most advanced in the world at the time, massive new power stations were being built.[12]

CHAPTER 15

BATTERSEA AND THE BARRAGE BALLOON

In the countryside, it was the march of the pylons that evoked Stephen Spender's 'quick perspective of the future'. In London it was the appearance on the south bank of the Thames of a giant brick-clad building with two huge chimneys that heralded a brave new world. When the residents of Chelsea got a sight of the plans they were horrified, as were King George V and a distinguished group of outraged citizens. This new power station at Battersea was to burn 2,000 tons of coal a day smack in the middle of London. The site had already been bought by the London Power Company, a consortium of electricity suppliers, and permission for it to be built had been granted by the Electricity Commissioners when *The Times* published a letter urging the government to reconsider.

Among the signatories were Lord Dawson of Penn, George V's doctor, the editor of the *Lancet*, the mayors of Chelsea and Westminster, and the president of the Royal Institute of British Architects. They believed there was still a chance to stop work on the power station going ahead. Their letter began: 'It appears to us that the schemes of the Central Electricity Board to erect super generating stations at Battersea and elsewhere in the midst of large towns have been framed without sufficient regard to the welfare of the community as a whole. We are particularly apprehensive in respect of the station at Battersea...'[1]

This was to be the largest coal-fired station built so far, and the experience with earlier versions had been unpromising, with widespread pollution from sulphurous fumes and dust. In Manchester it had been admitted that the Barton power station, burning 500 tons of coal a day, had caused serious damage to crops on farmland for miles around. Battersea would burn four times as much.

> As the prevailing winds in London are south-west, the normal flow of fumes from a station at Battersea would take a line over the Tate Gallery, Lambeth Palace, St James's Park, Westminster Abbey, the Houses of Parliament, St Thomas's Hospital, Whitehall, the National Gallery &c. while Battersea Park and Chelsea Royal Hospital are both closer to the site of the proposed station. The corrosion of the stonework of the Houses of Parliament is known to be due to sulphurous vapours: the fumes from Battersea will be far more corrosive than the present atmosphere.

The Times letter went on to point out that the riverside site was prone to fogs and mist, which would trap the emissions from the power station chimneys. It should not be built in the centre of a large city unless there were some means of cleaning the emissions before they were wafted into the atmosphere. If damage was caused, what would be the cost in terms of claims for such pollution? And the authors pointed to Germany, where the power stations were now generally sited on the coalfields and the electricity they produced was carried 300 miles to the cities along cables laid on the bed of the Rhine rather than on overhead cables. London had a coalfield close by in Kent, not sixty miles away, where a power station could be built. Prevailing winds would carry pollution out to sea, and the daily deliveries of coal on the Thames would not be necessary. Cables could be laid on the bed of the river to bring the power into central London.

> With these considerations in view and, above all, the danger to our national treasures, to the health and comfort of the nearby population and to the vegetation of our parks and open spaces, it seems manifest that the present project is ill-advised, particularly as there are alternatives which have technical and economic advantages.

The Cabinet considered these objections, which, on the face of it, seemed hard to refute. But it was unmoved. A memorandum from the Electricity Commissioners noted that in Paris a new power station had been built away from the centre of the city and that many were now being built in suburban areas, but that New York was getting a new city centre station. As far as London was concerned, the choice between a downriver power station supplying central London and the Battersea proposal was settled by economics. A new power station at, say, Dartford in Kent would cost £3 million more than the Battersea station when account was taken of the price of the cabling and losses of power sent over long distances. To get power from Kent to central London would require thirty-six separate 66,000-volt cables, which would have to be buried underground as there was no question of using overhead lines in a built-up area. The Port of London Authority, in charge of shipping on the Thames, ruled out the laying of cables on the riverbed.[2]

The Commissioners were confident that there would be no emissions of grit and soot from the power station chimneys, but were less sure about the means of getting rid of sulphurous fumes. All they could say by way of reassurance was that the London Power Company was taking the best advice available on how to deal with the problem. In short, every objection raised was countered and the Cabinet was happy to let the scheme go ahead. The building of the Grid had not saved London from a line of power stations all along the Thames: the state of technology in the 1930s was such that shipping coal from the mines to the centres of population was cheaper than building power stations on the coalfields and transmitting the current to the cities.

The overruling of the objections to Battersea arose from the same political will that brushed aside protests about pylons: the country needed cheap electricity in order to modernize, and nothing would stand in the way of that ambition.[3] However, as fate would have it, just as the demand for electricity was beginning to rise steeply in the late 1930s and the twin towers of Battersea power station appeared on the London landscape, the sirens began to wail and the capital was plunged into darkness at night.

The outbreak of war in September 1939 inevitably put an end to the rapid electrification of Britain promised by the completion of the first National Grid and the building of new and much larger power stations. In fact, in the year or so before war was declared there was anxiety that the nation had made itself more vulnerable than ever, with its pylons and towering power station chimneys that would be an easy target for enemy aircraft. Preparations to protect power stations and sub-stations had already begun in 1936, many of them protected with brick casing and camouflaged after aerial surveys to check their vulnerability.

In effect, the war turned the entire edifice of Britain's electricity supply system on its head. It had been in London and the south-east of England that demand had been growing rapidly, with the new factories and the growth of semi-detached suburbia which doubled the built-up area of the capital between 1918 and 1939. With the outbreak of war, Battersea power station was clearly now in the wrong place. The blackout and restrictions of the domestic use of electricity brought an immediate slump in domestic demand at the outbreak of war, and the site on the river was no longer an advantage. The collier ships that brought the daily ration of coal from the north down the east coast were too easy a target for the German U-boats, and the supply of coal for London had to be diverted to the railways and canals.

Before the outbreak of war, strenuous efforts were being made to keep to a minimum the pollution from the big power stations, especially those at Battersea and Fulham in London. However, when the bombing began in 1940 the Ministry of Home Security issued directions to increase the amount of smoke emitted from coal and oil furnaces, 'so as to intensify the general haze in the areas concerned with a view to hindering attacks of hostile aircraft'. This was not enough, however. At both Battersea and Fulham power stations 'gas washing' plants had been introduced to clean the sulphurous fumes that so concerned conservationists, who imagined the priceless art of the Tate Gallery across the river dissolving in an acidic haze. One effect of gas washing was to produce 'long and substantially horizontal plumes of condensed water' issuing from the tall power station chimneys. After power stations on the Thames were hit during the air raids in the autumn of 1940, the Air Ministry asked the Electricity Commissioners if these plumes, which were so easy for the Luftwaffe to identify, could be prevented. After a brief consultation with various government departments, the gas washing was suspended for the duration of the war and London got a dose of oxides of sulphur as well as a deliberately generated smoky power station haze.[4]

New munitions factories were established out to the west, away from the bombs that were expected to fall chiefly on London and the industrial areas of eastern England. The power stations on the Thames looked to be sitting ducks, their position neatly marked by the night-time glint of the river. Nevertheless they had to keep working, and the power they produced was siphoned off not for the brave new world of electric suburbia but for factories a long way from the capital. A central control room for the whole of the Grid had been established just before the war at Bankside power station. When this was hit it moved to an abandoned tube station just on the other side of the river by St Paul's Cathedral. The control centres in the other regions were all linked to London by telephone, so that supplies could be switched quickly if power sources were knocked out. The pressure of demand on this system grew enormously during the war, as existing suppliers tried to match the needs of the munitions industry. No large power stations could be built during the war – their size was limited by government decree – and the necessity of keeping demand down to conserve coal stocks became more and more pressing.

Because of the vulnerability of the Grid in London and the east of England, and the need for power in remoter western regions of the country, entirely new sections of Grid lines were built during the war. As all industrial output in Britain was concentrated on the war effort, the order for the steel towers and cables went to American firms. Between 1940 and 1941 the component parts of a wartime Grid were sent across the Atlantic in fifty-two shipments, running the gauntlet with the U-boats. All but two ships got through.

German bombers hit London's Fulham power station, Bankside and Battersea, but neither they, nor the V1 and V2 flying bombs that landed towards the end of the war, seriously threatened the supply of electricity. Only four pylons took direct hits, and few lines came down under enemy bombardment. In fact, during the five and a half years of the war, 'hostile action' gave rise to only 366 faults, 8 per cent of the total inflicted on the electricity supply system in those years. By far and away the most significant problem for the industry was 'defensive action'.

To protect London, harbours and many strategic positions, a Balloon Command force was organized before the outbreak of war. Tethered on wire cables, large unmanned airships were floated in lines to form a curtain across the expected line of attack of enemy aircraft. Any plane that struck a cable was likely to be brought down. And the barrage balloons were raised to a height that would prevent low-level strafing attacks while, at the same time, forcing bombers high enough for the anti-aircraft guns to have a chance of hitting them. They were also regarded as a good line of defence against low-flying German planes, which dropped mines around the coast and in the Thames estuary.

As early as 1936 the Committee of Imperial Defence ordered nearly 500 balloons for the protection of London. They were familiar enough by the outbreak of war in 1939 for Harrods to offer toy models as Christmas gifts for boys. By the time of the Battle of Britain in August 1940, there were well over 2,000 balloons ready for action. Each one was looked after by a ground crew that could haul it in for repairs – they were often shot at by enemy aircraft – or raise it to a required height. The balloons were designed to make life difficult for the Luftwaffe but they arguably caused more problems for the engineers trying to keep the electricity Grid in operation.

One of the first incidents happened in November 1939, when a number of balloons broke loose at once and drifted across the Thames estuary, their trailing cables severing power lines. Just before they drifted out to sea they cut right through the cables of the highest pylons that took the Grid across the Thames estuary. Where the environmentalists had failed, the barrage balloon triumphed, inflicting regular damage to the Grid. Their trailing cables were especially troublesome when aerial battles were fought over the Downs during the Battle of Britain in 1940.[5]

Of the 4,607 recorded major faults in the Grid system during the war, just under half were not attributable to hostilities. Of the remainder, more than 40 per cent were caused by defensive action, and nearly all of these – 1,614 incidents – were the result of free-flying barrage balloons.[6] They were especially destructive when they dragged their cables across country, slicing through one set of high-voltage cables after another. This was a nightmare for those in charge of the control rooms, who were trying to maintain power despite the outages. A post-war report highlighted one section of high-voltage transmission lines (it did not say where) which had temporary joints put in sixty-one out of its total of 102 spans as a result of damage by drifting barrage balloons.

The Grid, nevertheless, not only survived but proved to be invaluable during the war. As anticipated, the demand for power from the munitions industry rose steeply after 1939 while domestic consumption fell, and the pattern of 'maximum' load changed with the blackout. Provided no light showed, it was permissible to have lights on indoors, but the street and shop lights had to go. The public were given instructions on how much electricity various appliances consumed, and there was an order that they should not exceed 75 per cent of their use in the year before the outbreak of war. People with electric cookers were given an extra allowance. As a rough guide, the public were told that one kilowatt used for one hour equalled one unit. This amount of power would be sufficient for one of the following: it could run an electric clock for one year; provide five gallons of hot water; boil twelve pints of water; heat an electric iron for three hours; run a vacuum cleaner for six or seven hours; and light a 40-watt bulb for twenty-five hours or a 60-watt bulb for seventeen hours. It would only give one hour's warmth, however, from a one-kilowatt electric fire.

Until the outbreak of war the competition between gas and electricity had been fierce, with the rival industries doing their best to attract or keep domestic customers. At the outbreak of war they were faced with a Fuel and Lighting Order, which meant rationing. Both industries reacted angrily. The gas lobby said that the higher the rationing in the home, the more coal would be burned. The electricity lobby argued with the Mines Department that the blackout had already cut down domestic consumption of electricity and no rationing was needed.

However, the two sides recognized that they had a greater chance of fending off rationing if they decided to bury the hatchet for the duration of the war. Neither would try to poach customers from the other, nor would they promote the use of gas or electricity. This 'truce' naturally put an organization like the Electrical Association for Women in a difficult position, as their raison d'être was the promotion of the use of electricity in the home. Caroline Haslett, the director of the EAW, had some amusing exchanges with managers of electricity supply companies over the issue of illicit gas promotions and objections to EAW activities. Her wartime correspondence contains a number of stories of misbehaving gas companies trying to stop women getting electric ovens, and in one case adopting what she called 'Nazi methods'.[7]

The truce held pretty well though, and the EAW found other activities to help in the war effort, one of which was a project for mobile canteens set up jointly with the Young Men's Christian Association. These canteens provided, as well as cups of tea, a good selection of chocolates, cigarettes of various brands in packs from five to twenty, and chewing-gum. For the men away from home there were razors, shaving soap, toothpaste, Brylcreem, boot polish and pencils. In addition the EAW created a 'Cigarette Fund', which would allow them to offer each soldier stationed in England one cigarette for every visit to the canteen. They also arranged with Women's Institutes and other bodies to knit and supply socks, collect books, games and pencils and notepaper. Despite the displacement activities of the EAW, and the uneasy truce with gas, domestic consumption of electricity began to rise after 1942 and it was higher than in 1939. It was the winter peaks which brought this about, and it became clear that when coal was rationed in January 1942 those who had electric fires made much greater use of them. There was no effective way of rationing domestic electric supplies, as the power could not be stored and handed out in sacks like coal. When the end of the war came, the electricity industry was working at full stretch: it had expanded, but at nothing like the rate it would have done in peacetime. To bring on the electrical millennium, new power stations would have to be built and new Grid lines stretched across the country to carry higher voltages. And, crucially, coal stocks for the power stations would have to be replenished as miners returned to their pits and production could increase once again.

The victory of the Labour Party at the general election in July 1945 would have a profound effect on the future of the electricity industry. Since the very early days much of it had been in public ownership, run by local authorities. Labour's plan was to nationalize the whole of it, so it would be a state-run rather than municipal enterprise. At the same time the coal mines, on which electricity was entirely dependent, would be nationalized. The government would be in charge, and finally, after half a century, electrical power would be provided by and for the people.

One man who had dedicated most of his life to making electricity available to all was cruelly deprived of the satisfaction of looking forward to a post-war renaissance for the industry. On the night of 14 October 1940, at the height of the Blitz, the home of Charles Merz in Melbury Road, Kensington, was completely demolished by a German bomb. He died along with his two children and two servants. His wife, Stella, was injured but escaped with her life. Four years earlier, William McLellan, Merz's partner, had died. The great pioneers of mass-produced electricity were all gone by the end of the Second World War, when Britain embarked on a bold experiment which, in theory at least, would at last bring power to the people.

PART THREE

POWER FOR THE PEOPLE 1948–89

CHAPTER 16

THE SNOW BLITZ

If there was one thing the newly elected Labour government might have wished for in the first years of its ambitious programme of nationalization, it was a run of warm winters. The country was absolutely dependent on coal for its power and for domestic heating and cooking. In 1945, and for some years afterwards, the majority of people still relied on coal fires to keep warm in winter and many still had a coal range in the kitchen for cooking. Domestic consumers burned nearly 30 million tons of coal annually. That was more than industry, which took around 26 million tons, gasworks with 21 million tons and electricity power stations with 23.5 million tons. Yet, at the end of the war, coal stocks were at a dangerously low level and it would take time to get the mines back to full working as men returned from the services. And there was no alternative to coal – it provided 93 per cent of the nation's energy – no oil of any significance, nor much hydro-electric or wind power to generate electricity, nor any natural gas. Deliveries of coal from mine to factories, gasworks, power stations and homes were vital.[1]

The winter of 1946–7 did not start out too badly. On 1 January the coal miners celebrated the achievement of their long-held ambition to become state employees when the pits were nationalized. Manny Shinwell, Labour's Minister of Fuel and Power, for a while entertained the romantic idea that from that time on the miners would be in such good spirits that a fuel crisis that others saw coming might be averted. At the same time the government, intent on pursuing its policy of taking major industries into public ownership, was working on the detail for the nationalization of electricity. The Bill was published on Friday 10 January, setting out the terms of compensation for private companies and municipal suppliers and the structure of state ownership. Though some of the more powerful private electricity companies tried to put up a fight, the nationalization of electricity was not especially contentious.

There had been heavy snowstorms at the beginning of January but then the weather had cleared, so that when Parliament began to debate the future of electricity there were some quite balmy days. Temperatures reached 57°F (14°C) in places, just what Labour needed to conserve the diminished stocks of coal. A glance at the weather forecast did not suggest that there was trouble ahead, even on 23 January when it had become colder and there was a warning of 'scattered wintry showers'.

In *The Bleak Midwinter*, his definitive account of the astonishing cold spell that was to follow, and which some newspapers called the Snow Blitz, Alex Robertson wrote:

> In the evening of the 23 January it began to snow. This was no slight spattering from 'scattered wintry showers' but a steady, silent downpour which enveloped London and the south-eastern counties. By the morning of Friday 24th the south-east was blanketed to a depth of as much as six inches... More snow was

> forecast for the weekend, and duly arrived, accompanied by intense cold... The Saturday was not too bad; cold and with flurries of snow, but relieved by occasional intervals of sunshine. Sunday 26 January was something else: *The Times*, not usually given to over-dramatising the news, felt justified in describing it as 'one of the wildest days of the winter, with frost, snow, and a bitter wind that sometimes reached gale force.[2]

The Sunday blizzard which had begun in the south-east moved northwards covering the whole country, forming deep drifts in many places. There was four feet of snow in the countryside. And there was no let-up. The freeze lasted for seven weeks: Kew Observatory recorded no sunshine from 2 to 22 February. Bitter east winds kept the snow crisp, and it covered the ground for most of England and Wales from 25 January until the middle of March. One blizzard followed another, and the newspapers carried pictures of steam trains completely buried under drifts. The RAF had to deliver food to isolated towns and villages, some of them in mining areas that could produce no coal.

As women began to queue with prams at gasworks around the country in the hope of taking home a bag or two of coke, the power cuts began. At first they were unpredictable, then routine, with a regular timetable. No electricity to be used in the home between nine in the morning and midday and between two and four in the afternoon. Those who had been seduced by the ideal of the 'all-electric house' just before the war now found themselves cut off from their only fuel for cooking and heating. Power stations could not keep up with the demands of industry, and more and more people were thrown out of work because factories could not open. To conserve power the BBC Third Programme closed down, and greyhound racing was banned as well as midweek football matches. Cinemas, theatres and music halls had restricted hours. The Cabinet met by candlelight on many occasions, and at one point contemplated the awful possibility that there might be famine. Farm workers were pictured digging turnips with pneumatic drills, the ground was frozen so hard.

When the thaw came in mid-March there was widespread flooding, but the summer that followed was hot and sunny and the horrors of the winter of 1947 began to fade. As they did so the expectation of a return to some kind of normality began to rise. But what was 'normality' as far as electricity was concerned? One of the supposed benefits of nationalization of the industry was to be a more widely available supply of 'cheap electricity'. State ownership would get rid of all the inefficient little stations and remove once and for all the obstructive 'parochialism' of the supply industry. It was the same kind of argument that had brought in partial government control and the Grid in the 1920s and would now, finally, get rid of the legacy of Victorian Liberalism which had been responsible for the fragmented nature of the British industry.

The British people certainly appeared to be ready to adopt electricity in the home at the end of the war. At Christmas 1946 families had allowed themselves a decent celebration for the first time in seven years. On New Year's Day 1947 *The Times* ran a piece with the headlines:

Cabinet and Coal
Hope of Avoiding Domestic Cut
Electricity Demand up 38 per cent

The report noted: 'A striking feature of Christmas week was the heavy consumption of coal for electricity and gas in spite of the considerable reduction in industrial demand. In the week ended December 28, 596,000 tons of coal was consumed in electrical generation – an increase of 15.3 per cent over 1945. The rise in the amount of coal used for electricity is specially significant as indicating the very great expansion in domestic demand.'

Now that the wartime truce with gas was over, the electricity industry was in a position once again to promote the great advantages of its clean and efficient power and rapidly multiplying range of gadgets available for the home. But there was a problem. Until new power stations were built there would simply not be the supply to cope with an increase in domestic demand, which had risen threefold between 1939 and 1948. It was one appliance in particular that had boosted the domestic use of power: the electric fire. These had been around since the early 1900s, but it was the modern design of Ferranti that had made the electric fire especially efficient and popular. In 1929 he came up with an arrangement whereby a polished metal parabolic reflector threw out the heat from a glowing wire element. There were models that could be fixed in an old fireplace as well as cheaper portable versions. Both had a pleasing glow to them, a comforting visual display of warmth that some earlier electric fires lacked.

The rise in the household consumption of electricity towards the end of the war had been attributed chiefly to the use of electric fires when coal was severely rationed. Now, in peacetime, it was the popularity of these fires that pushed up demand for power even further. They were guzzlers of electricity and had to be used sparingly if the electricity bill was to be kept under control. But, in the immediate post-war years, electricity no longer appeared to be all that expensive. In fact, just as the industry was nationalized, there was a strong argument to be made that it was now *too cheap*! It was calculated that in 1948 the average domestic consumer was using twice as much electricity as ten years earlier and paying, in real terms, only half as much.[3]

One of the causes of the low price of electricity in the immediate post-war period was an unintended consequence of nationalization. The owners of the commercial supply companies that were taken over by the state were given compensation, which was regarded as reasonable and cost £542 million. However, the larger part of the supply industry had been owned and run by local councils, and the Labour government did not consider them eligible for compensation in the same way. Many were aggrieved that they were losing their autonomy as municipal traders. In the year or so before nationalization they saw no reason to raise the price of electricity and preferred to give their customers, most of whom were ratepayers, an easy time, while, as suppliers, they went into the red.

The British Electricity Authority, which took over from the Central Electricity Board in April 1948, inherited about £7 million worth of municipal debts and some unrealistically low tariffs. When these were adjusted upwards there was an outcry and the inevitable headline from the *Evening Standard*: 'You own the electricity industry – and up go the prices!' Raising the price of electricity was one very obvious way of curbing demand, but in the immediate post-war years this was political dynamite. In 1948 a committee chaired by Sir Andrew Clow, a former governor of Assam, had recommended a scheme in which electricity was charged at a higher rate per unit in winter than in summer. If the electric fire factor could be controlled on chilly mornings, when national demand peaked, there would be fewer outages.

The chairman of the newly created British Electricity Authority was Lord Citrine, who at first agreed to the Clow tariff plan which was due to be introduced just as nationalization took effect. However the area managers who would have to implement it and face the public put up a staunch opposition. Citrine did his best to find a compromise. Walter Citrine was born in 1887, the son of a Liverpool seaman, and had made his way in the world without any formal education. Leaving school at the age of twelve, he became an apprentice electrician and an early member of the Electrical Trades Union. Through his union work he became a familiar figure moving in the highest political circles when he became General Secretary of the Trades Union Congress.

Citrine had a well-deserved reputation as a tough but fair negotiator, and in the end found agreement on the 'Clow tariff' with a winter charge a little above that of summer. It had no appreciable effect, and the problem of excess demand remained. In a memoir, Citrine recalled this period: '...we had to urge consumers, not only to go easy at peak times but at other times also. I became so economy minded that when I walked into a restaurant or a hotel it made me indignant to find all the lights blazing away in broad daylight. A fine state of affairs for a Chairman whose job was to sell electricity!'[4]

There was to be no quick fix for the electrical industry, and one of the first tasks for the British Electricity Authority was to run a campaign urging families not to use their few electrical appliances at peak times. 'Who ruined breakfast?' was one slogan. Four cartoon heads looked on smiling: Mr Hancock, Old Mr Rawlins, Mrs Osborn and Young Bob Cross. The story read: 'A pleasant breakfast in the comfort of the home can be marred unless we are careful with electricity during the Peak Hours. Lack of thought on the part of a few may mean discomfort for many. We *must* go easy with electricity during the Peak Hours. If too many people try to use too much electricity during Peaks, the power stations have to cut supplies, and that means discomfort and inconvenience. We can plan our day to use the electricity we want in *Off-Peak* hours.' Exactly when people were supposed to have breakfast was not made clear, but there was an explanatory box:

THE PEAK – What it means
Power Stations have 'Peak Hours' –
just as buses and trains have rush hours. They
occur when, in addition to the essential
needs of industry, the demand for electricity
from homes, offices and shops becomes too heavy.
From Monday to Friday the usual Peak
Hours are: **8.0 to 9.30 a.m.**
from Oct 1st to March 31st
and in addition:
4.0 to 5.30 p.m.
from Nov 1st to Jan 15th
During these times the *factories must come first*

A debate arose in the pages of *The Times* about the wisdom and efficacy of 'load spreading'. The Labour MP and former Minister of Fuel and Power Philip Noel-Baker wrote in March 1952 to point out that there had been many fewer power cuts in the winter than in 1950–51, and one of the reasons was that load-spreading had worked.[5] Factories, shops, cinemas and housewives had all got into the spirit of the thing. An Area Board could even phone a department store when the supply looked like being cut off and ask for the power to be turned off for an hour or two. The demand for electricity had gone up 37 per cent in the past four years, and it could only be dealt with by load-spreading.

However, the flamboyant Tory MP Gerald Nabarro, famed for his handlebar moustache and military bearing, disagreed with Noel-Baker, describing load-spreading as 'at best a mere palliative, not a policy'. The startling solutions put forward by Nabarro included a return to coal fires in the home. 'Throughout the post-war years everybody has been starved of solid fuel at home. So the householder buys an electric fire, plugs in and burns his coal at the power-station instead. That is, largely, why electricity consumption has risen so greatly since the end of the war. But the effects of this homely switch upon national resources are extremely damaging...'

According to Nabarro's calculations (which were quickly disputed in *Times* correspondence), heating a room with an electric fire used twice as much coal as burning solid fuel in a grate or extracting gas from coal to burn in a domestic fire. It was the electric fire that put pressure on the whole system, so that expensive equipment had to be installed in power stations to deal with peak loads. This would be dealt with in Nabarro's first fuel policy initiative: '...the electric space-heating load should be coerced and persuaded on to solid-fuel and gas heating... and the consumer given the compensatory fuel to allow him to abate the grid demand.' Part two of Nabarro's strategy for fuel and power was to encourage factories to have their own generators, which would lift pressure from the Grid. In other words, the Nabarro strategy was to turn the clock back to a time when electricity was little used and locally generated.[6]

While the debate about what to do about the shortfall of electricity supply rumbled on, the British Electricity Authority began ambitious plans to overhaul the entire electricity supply system, which was evidently already out of date. The National Grid, built in the 1930s and added to in the years of the war, could not carry the loads the nation now needed with its ever-rising demand for electricity. Boldly, the BEA decided to build what became known as the 'super-grid', with much taller towers and cables that could carry twice the voltage of the old Grid. The wayleave officers now had to contend with a new Town and Country Planning Act when siting the pylons, which had a standard height of 136 feet 6 inches, 'five times as tall as a house', as the newspapers put it.[7]

The completion of the Super-Grid brought about a fundamental change in the economics of electricity supply. When the first power station at Battersea was built, it was argued that it was cheaper to ship coal up the Thames than to lay cables from a more distant station out of town. Once the 275kV cables were in service, it was cheaper to transmit electricity long distances than to transport coal. New coal-fired power stations were therefore built in the mining areas, and oil-fired stations, when they came in, could be built next to refineries. After the war Battersea was completed with its third and fourth towers, which gave it the familiar appearance of an up-turned Art-Deco table, but really by then it was close to being an anachronism.

In December 1952 the weather once again played a part in concentrating the minds of those debating the solution to the problem of electricity supply. On Thursday 4 December a high pressure area settled over London, the breeze of the previous days dropped, and a fog descended. From thousands of chimneys smoke rose and mingled with the moist mist of winter. Fog became smog, so that it was impossible for anyone to see where they were going. Buses and trains could not run. Handkerchiefs held to the mouth soon turned black with soot. People suffering from breathing difficulties began to fill hospital beds and the number of deaths rose rapidly as the smog hung on, right through the weekend and to the following Monday. The choking, blinding pall in London did not begin to lift until the Tuesday.[8]

For many years there had been a Smoke Abatement Society – William Siemens belonged to it in the nineteenth century. Those who advocated the use of electricity for the home argued that it would make the air in cities cleaner. But the coal fire remained popular, and in London and other cities the steam railways and factories and power stations continued to pollute the winter air in a manner which is unimaginable today. The Great Smog of December 1952 was a turning point. A committee chaired by Hugh Beaver, an engineer, was appointed in 1953 to examine the problem of air pollution and recommended the following year that the government should bring in legislation to create 'smoke control zones', in which pollution would be minimized.[9]

When the government dragged its feet over implementation of Beaver's recommendations, a private members' Bill was introduced in the Commons by, of all people, Gerald Nabarro, who had not long before advocated a return to coal fires. A Clean Air Act was finally passed in 1956, which, within a few years, outlawed the burning of anything other than smokeless fuel in an open fire. Grates designed to burn house coal were no good for coke, however, and the open fire rapidly disappeared from British towns. This hastened the switch to gas and electric heating and within six years or so put an end to smog in London and other major cities.

However, Britain remained absolutely dependent on coal to produce electricity, and it became clear in the early 1950s that the demand for power would rapidly outstrip the supply of coal from the pits. The simplest solution was to import coal to make up the shortfall on home production. But there was a recognition that in the long run there would have to be other ways of generating electricity. Demand was set to rise dramatically as more and more pieces of electrical equipment became standard in British households. Between 1945 and 1965 the nation at last became electrified, not quite as thoroughly as the United States but recognizably 'modernized'. Nearly everyone, 94 per cent, had an electric iron by 1965; 81 per cent had a vacuum cleaner, 85 per cent had a TV set, 70 per cent had at least one electric fire, nearly 60 per cent had a washing machine, 48 per cent had a refrigerator, and more than a third of households had an electric cooker. Microwaves, dishwashers and food mixers were still very much a novelty but would soon put even further demands on the electric supply industry.[10]

There were alternatives to coal. Oil could be burned to heat the boilers for the steam turbines, but it was expensive. The mammoth Bankside station, built in 1947 on the site of an earlier power station, was oil-fired but the experiment was not a success. There was, too, the prospect of generating electricity with the waste heat from a nuclear power station. But that was futuristic in the early 1950s, the technology shrouded in secrecy. A more promising prospect, in the short term, was the possibility of making use of Britain's untapped energy in the Highlands of Scotland. The wild north had been excluded from the mainstream of electricity development in Britain, and was a region free of pylons and Grid cables left to its own devices. Before the National Grid was built there was little prospect of harnessing the power of the tumbling rivers and lochs of the Highlands. Now, however, if the sites could be found and the environmental issues resolved, Britain might get some valuable hydro-electric power to top up the Grid.

CHAPTER 17

POWER FROM THE GLENS

In the late nineteenth century, when much of Britain had been transformed into a soot-blackened industrial landscape, the glens and tumbling tea-brown rivers of the northern Highlands of Scotland were cherished as a wilderness, a picturesque and romantic survival of natural beauty. The notion that this landscape had been untouched by commerce and industry was fanciful: the expulsion of the crofters to make way for sheep farming had been a traumatic period in the region's history, and it had become the playground for rich industrialists who liked to stalk deer and cast a fly for salmon. Nonetheless, there was a rugged beauty in the Highlands that was not matched in the English Lake District or the mountains of Wales and which attracted tourists, who gaped at the scenery celebrated in romantic literature and poetry.

Waterfalls were especially attractive to the Victorian sightseer; the rush of foaming surf over rock was exciting, and redolent of untamed natural forces. The Falls of Foyers was one such, a favourite on the MacBrayne steamer tours of Loch Ness and the Caledonian Canal. It could not compare with the world's great waterfalls, fed as it was by a modest river that tumbled down into the eastern shore of Loch Ness. But when it was in spate, the roar and spray in the magical setting of rugged scenery as the white water fell 165 feet to the loch was mesmerizing. Robert Burns penned a verse in pencil while admiring the Falls in 1787:[1]

> Among the heathy hills and ragged woods
> The roaring Foyers pours his mossy floods;
> Till full he dashes on the rocky mounds,
> Where, through a shapeless breach, his stream resounds,
> As high in air the bursting torrents flow,
> As deep-recoiling surges foam below,
> Prone down the rock the whitening sheet descends,
> And viewless Echo's ear, astonish'd rends.
> Dim seen, through rising mists and ceaseless showers,
> The hoary cavern, wide-surrounding, lowers.
> Still, through the gap the struggling river toils,
> And still, below, the horrid cauldron boils.

It was the power of this 'horrid cauldron' which in 1894 caught the eye of prospectors working for the newly formed British Aluminium Company Ltd. It had been known for many years that the lustrous metal aluminium (or aluminum as it was called in the United States) could be extracted from specific rocks by an electrolytic process. Humphry Davy had demonstrated the procedure as early as 1806. But there was no commercial way of exploiting it until the development of electric generators. Even then, with the first Gramme machines, not much aluminium was made and it remained a semi-precious metal until the 1870s. The breakthrough had come in 1866, when a process that could mass-produce aluminium was discovered simultaneously by two young inventors, both in their early twenties: Charles Hall in America and Paul Heroult in France.

The British Aluminium Company Ltd, with the eminent electrician Lord Kelvin on the board, was established to exploit the French patent. Bauxite (named after the small town of Les Baux, where it was first identified), from which aluminium could be extracted, was plentiful in Antrim, Northern Ireland. The most important decision the company had to make was how it would generate the large quantities of electricity needed for its works. Its decision to go for hydro-electric power was an economic one. A coal-fired power station might be cheaper to build but the running costs would be far too high. The price of aluminium was falling rapidly as the Hall-Heroult process was adopted in America and Europe. In the long run hydro-electric power would provide inexpensive current. A site in Scotland close to the Caledonian Canal, which could transport heavy goods, seemed to be the ideal. And the celebrated beauty spot of the Falls of Foyers promised ample power to drive a generator.

To make use of the water-power potential of the Foyers river, it was necessary to divert the main flow of the stream above the falls into a large pipe that would deliver the thrust to turn the generators. Crudely speaking, the power of a hydro-electric system can be determined by the quantity of water multiplied by the height it falls. The amount of water available could not be left to chance. Above the Falls of Foyers there would have to be a reservoir with a dam to ensure supplies even during a dry spell, rare but by no means unknown in the Highlands. Surveying the pristine scene on the eastern shores of Loch Ness, the Aluminium Company's engineers could see that joining two small hill lochs above the falls and damming one end would be all that was needed to create a reliable hydro-electric power plant.

However, when the plans became known there was a howl of protest from lovers of the Highlands. In August 1895 *The Times* published a venomous letter from the Duke of Westminster, then president of the National Trust for Places of Historic Interest or Natural Beauty:

'Workmen are at present employed at Foyers in making a tunnel from a point above the upper fall to a point below the lower fall. Through this tunnel the whole of the waters of the Foyers river will be conducted for the purpose of manufacturing aluminium by means of electricity, so that, as the agent of the Aluminium Company states, "the falls will not be injured, only there will be no water in them."'[2]

The company denied that any representative had suggested the falls would disappear altogether, but in a counter to the Duke's claim stated in a letter from the secretary, published in *The Times* on 9 October: 'The company has never contemplated taking all the water passing over the Falls, or interfering with their natural structure, and such water as the company do not use will pass over the Falls as at present, but should the company's proposals for the storage of water during flood times not be allowed, the company may reluctantly be compelled to utilize in times of scarcity the whole of the flow.'

Among those who added his authority to the protests was John Ruskin, his opinion that the desecration of the falls would be an 'iniquity' recorded in a letter sent to *The Times* by his cousin and helper Joan Ruskin Severn. She had read Ruskin a published letter from the secretary of the National Trust, H. D. Rawnsley, in which he pleaded: '...surely for the sake of the nation, which each year more clearly recognises the worth of natural beauty to its national life, and even from the lower utilitarian motive that year by year more people are tempted to visit Scotland because of such attractions as Foyers Waterfall affords, the county council should be urgently called upon to rouse itself and spare no efforts to save so beautiful a piece of scenery.'[3]

M. J. B. Baddeley, editor of the *Thorough Guide* series for tourists, asked: 'Can nothing be done to save this magnificent bit of British scenery – the *bonne bouche* of the Caledonian Canal route?... Their scheme is the greatest outrage on Nature perpetrated in the present century, and the excuses made for it are the most inexcusable.'[4] Baddeley was e specially angry with an engineer, not associated with the company, who had written to say that perhaps the falls could be turned on and off and illuminated at night with the electricity generated for the amusement of tourists, as was, apparently, the practice in parts of Switzerland.

All the protest was of no avail. The company had taken the precaution of buying 8,000 acres of Highland moor, loch and hill as well as the accompanying water rights, and had only to negotiate with Inverness County Council over the diversion of roads.[5] The Falls of Foyers – there are in fact two as the river tumbles down to Loch Ness – lost some of their grandeur but never quite dried up. The British Aluminium Company expanded as the demand for their product rose steeply in the early twentieth century, and they were soon building a second hydro-electric station further down the eastern bank of Loch Ness at Kinlochleven. For this second power station 3,000 workmen were encamped alongside Loch Ness to build a dam 3,000 feet long and ninety feet high, the largest structure of its kind in Europe at the time. Begun in 1905, it was one of the last great works of the so-called 'navvies' whose muscle had cut the canals from the eighteenth century and laid the railways in the nineteenth. The spectacle of this rough encampment in the Highland landscape was yet another blow for the conservationists.

The invention of various industrial processes that required large amounts of electricity continued to make the Highlands an attractive region for manufacturing companies. In the early days of the development of electricity in the north of Scotland, that was the only purpose for which large hydro stations were built. Exploiting the hill lochs, rushing burns and frothing waterfalls for domestic supplies was quite another matter. The Highlands were too remote and too sparsely populated for any commercial developments other than those associated with industry to attract investors, and the whole northern region was regarded as beyond the Pale.

Here and there in the Highlands, however, there were very early schemes that used water power to generate electricity for lighting. In 1890 the monks of St Benedict's Abbey in Fort Augustus, at the south-western tip of Loch Ness, worked a turbine from one of the streams running into their grounds and were able to generate enough to light the village. Fort William on the west coast was lit from a hydro-electric station built in 1896, and there were several other small-scale schemes before the Great War. But most of the Highlands remained without electric power well into the twentieth century.[6]

The trauma of the desecration to the Highland scene that began with the Foyers scheme gave hydro-electricity a bad name in Britain, and for many years to come it was regarded as essentially destructive rather than as a clean alternative to coal. And yet it was recognized by various government committees that looked into the potential for hydro-electricity that 80 per cent of the promising sites were in northern Scotland, where there were high lochs that could become storage reservoirs and steep falls of water to drive generators. The only other region with much promise was mountainous North Wales.

However, hydro-electric power, along the lines of the pioneer scheme at Godalming in Surrey in 1881, was not entirely ignored outside the Highlands. In fact there were many small-scale water-powered generators at work in England. The steam engine had made watermills obsolete and many lay unused for years. It occurred to a few enterprising electricity supply companies that they could harness redundant mills to generators to supplement their main supply from a coal-fired station. From June 1918 until 1921, a Water Power Resources Committee was set up by the Board of Trade to investigate the extent of untapped energy in the rivers that might be of use to industry. The chairman was Sir John Snell, an electrical engineer who had worked with Rookes Crompton in the 1880s and had had experience as a municipal engineer in London and the north-east of England before establishing himself as a consultant.

Though it was clear that hydro-electric power would never have the importance in Britain it had in Norway, Canada, Switzerland or the French and Italian Alps, the Snell Committee made a strong case for its use to help conserve coal stocks. It highlighted schemes in Chester and Worcester where once-redundant weirs had been revived to work turbines and save the cost of a few tons of coal a year. An attempt was made to estimate the untapped potential of rivers and lakes throughout the country, with rainfall figures collected from enthusiastic amateurs and engineers engaged to sketch out potential hydro-electric schemes. Though there appeared to be some scope in England for greater use of hydro stations, and considerably more in North Wales, where there were already industrial plants like those on Loch Ness, it was obvious that it was in the Highlands of Scotland that conditions were most favourable.

There was one scheme created below the Highland line that demonstrated a use of hydro-electric power along the lines suggested by the Snell Committee.[7] Southern Scotland had plenty of coal to fuel its power stations but there was always a concern that stocks might run low at some time in the future and that an alternative source of power would be valuable. The largest generator of electricity in Scotland just after the end of the 1914–18 war was the Clyde Valley Electric Power Company, with three large coal-fired power stations. This company appointed an engineer, Edward MacColl, who had gained a reputation for efficiency and ingenuity when he had run the Glasgow tram system. With no interest in hydro-electricity initially, MacColl was asked to find a way to harness the power of the Falls of Clyde, a beauty spot beloved of poets and as famous as the Falls of Foyers. The initiative came from a consortium calling itself the Power and Traction Finance Company, which would work within the Clyde Valley Company's area.

MacColl devised a scheme to tap the flow of water in the Clyde as it fell through four falls, without the need for any reservoir at the head. A series of sluice gates siphoned off the river water, which was channelled into pipes that turned turbines in two separate power stations before returning to the main river. The huge catchment area of the Clyde, which ensured a strong flow of water at all seasons, made this possible. And the picturesque falls were preserved. However, the Falls of Clyde hydro scheme was always an adjunct to the coal-fired power stations, providing a saving on solid fuel.[8]

There were many other proposals for schemes to tap water power in southern Scotland, but for the most part they made no economic sense and always threatened to unleash a storm of protest because they involved the damming of rivers and raising the level of natural water in the lochs, which were to act as stores of power. Saving coal was not a sufficient reason to embark on huge engineering works, and there appeared to be no other logic for building hydro-electric power stations in southern Scotland. However, the building of the National Grid brought about a dramatic change. The northern Highlands were excluded, but southern Scotland would now be linked by pylon and high-power cable to a much larger region and hydro-electric stations would be able to sell some of what they generated to the Central Electricity Board.

The possibilities presented by the Grid revived a scheme to generate power in the Galloway hills that had lain dormant for years. It was a huge engineering operation which was promoted by, among others, William McLellan, the partner of Charles Merz who had been so influential in the creation of the Grid. Even more taxing than the technical problems was the political opposition to the scheme from other vested interests, not least the Mining Association of Great Britain. Parliamentary approval had to be granted, and there was a long fight before the Bill for the Galloway scheme was finally passed in 1929. One of the conditions was that 'all reasonable regard shall be paid to the preservation, as well for the public as for private owners, of the beauty of the scenery' wherever works were carried out. The scheme was anyway delayed for two years while the Galloway Water Power Company awaited the decision of the Central Electricity Board to make it a selected supplier. That was where 90 per cent of the power it generated would go.

It was October 1936 before the scheme was in operation. Sadly William McLellan did not live to see the power turned on: he died in 1934 at the age of sixty, worn out, it was said, by the battles to get the Galloway project accepted. This was the largest hydro-electric scheme in Britain at the time, with five stations powered by a cascade of water as it passed from the highest loch to sea level. To allow salmon to continue their run to the spawning grounds on the upper River Dee and its tributaries, a special ladder was constructed at Tongland near the coast.

The electricity from the Galloway scheme was immediately in demand by the CEB and began to demonstrate the great value it had within the National Grid system. Whereas it takes days to get a coal-fired station up and running, hydro-electric power could be supplied instantly since the turbines were running continuously. This was to be the great contribution that hydro-electric power could make in Britain: stations in Scotland and Wales could be called upon at a moment's notice to meet peak loads, especially those surges in demand that were not entirely predictable.

Although the building of the National Grid held out the promise that quite remote power stations could find a market, this really had little bearing on the northern Highlands of Scotland. There was one major hydro-electric scheme in the region, the Grampian Electric Supply Company, which drew its water from Lochs Rannoch and Tummel and which started operations in 1930. It was near enough to connect to the Grid. But this really was not feasible for schemes further north.

The position of the northern Highlands beyond the reach of the Grid was little changed from the late nineteenth century when war broke out in September 1939. There was no prospect of a domestic electric supply reaching local communities, and the opposition to proposed schemes had been vociferous. In particular a scheme at Glen Affric, in one of the remotest regions of the Highlands, had failed to get through Parliament. The war would inevitably put an end to any new schemes, just as it brought a halt to the power station plans of the Central Electricity Board in England and Wales.

However, by a strange turn of fate the war transformed the official view of the potential for hydro-electric power in the Highlands. In 1941 Winston Churchill persuaded the distinguished Labour journalist and politician Tom Johnston, a Scot from Dumfermline, to become Secretary of State for Scotland. Johnston took the job, so it is said, on condition that he would be able to do something for the ailing economy of the Highlands, which were losing population steadily. Churchill's response was to say that if Johnston got together a group of past Secretaries of State for Scotland and they agreed on issues to be addressed, he would accept the outcome. The result was a determination to do something about the development of hydro-electric power in the Highlands.

Johnston assembled a committee chaired by the eminent Scottish lawyer Thomas Cooper, who had just been appointed Lord Justice-Clerk of Scotland. Reviewing earlier committees and drawing on the evidence in the Water Power Resources Committee, Lord Cooper cut through the history of hydro-electricity in the north of Scotland with a surgical precision and produced one of the most succinct reports ever to be delivered to a Minister. In contrast to the great developments of hydro-electricity abroad, the experience of the Highlands had not been 'inspiring'. Once again it was politics getting in the way of electricity:

> All major issues of policy, both national and local, have tended to become completely submerged in the conflict of contending sectional interests. Local opinion has vacillated widely from time to time and has been, and is, acutely divided both on general principles and on the merits of local proposals... The opposition of land-owning and sporting interests seems to have been pressed further than was justifiable for the protection of their interests and in some cases to have been taken more seriously than was intended; and in several instances strenuous opposition has been offered by the Mining Association with the object of preventing hydro-electric development so as to maintain or increase the demand for coal, though coal is virtually non-existent north of Fifeshire. In the result the whole subject has become involved in an atmosphere of grievance, suspicion, prejudice and embittered controversy.[9]

Lord Cooper and his committee had little sympathy for the opposition to hydro-electric power installations on the grounds of 'amenity'. A rise in the water level of a loch that is dammed to create storage does not necessarily harm the landscape, and the shores of these remote stretches of water were usually just rubble and useless as agricultural land. And if there was some alteration to the landscape, who was there to get upset about it? 'In the majority of cases,' states the report,

> the undeveloped water resources of the Highlands are situated in the loneliest and most inaccessible parts of the British Isles. We venture to doubt whether many of them have been visited once in a lifetime by one person in a thousand of the population of the United Kingdom.
>
> To take an example which has been much discussed, in the whole of the Glen Affric (over 12 miles in length) there are only 7 houses with an aggregate permanent population of 23 persons, consisting of 5 deer-stalkers and 2 gardeners and their respective households. In the next valley, Glen Cannich, there are 15 permanent inhabitants – stalkers, game-keepers and a roadman, with their families... What is true of Glen Affric is true to a much greater extent of most of the resources of the farther West and North. The only persons who have seen many of them in their natural state and who would be able to praise or to deplore the changes which would be affected by their development would be a handful of deer-stalkers, salmon anglers, ghillies and gamekeepers, and the adventurous spirits who have traversed the mountain districts on foot.[10]

Cooper identified what he called in his report 'two contrasting theories as to the future of the Northern Area'. One view was that the Highlanders should be left to carry on with their rustic crofts and coastal fishing and that they might be offered a supply of electricity to brighten up their daily lives. The other view was that hydro-electric power could be used to revitalize the economy of the Highlands and stem the outflow of young people in search of a more prosperous way of life. There was no doubt which 'theory' Cooper favoured. Hydro-electric power could provide the basis for an industrial renaissance, with particular industries in search of cheap power attracted to the region just as the British Aluminium Company had been in the 1890s. The alternative, in which crofters simply incorporated electricity into their everyday lives, was dismissed as totally unrealistic. It was the problem of rural electrification in an extreme form: a tiny, scattered population that could never give any hydro-electric scheme a return on its investment.

In fact the commercial development of the Highlands, in the first instance, appeared to be out of the question. What Cooper recommended, and the wartime Parliament accepted, was the creation of a public service corporation, to be called the North Scotland Hydro-Electric Board, which would take over responsibility for all schemes in the Northern Area. The three central tasks of the Board would be to provide cheap electricity with which to entice certain electro-chemical industries to the region, to assist existing power companies to expand with a view to selling surpluses to the Grid and to undertake some experimental installations in isolated districts.

The Cooper Committee more or less waved aside the concerns about the potential damage to the landscape of the Highlands, though when the subsequent Bill to create the new Hydro Board went through the House of Lords some prescient warnings were given even by its supporters. Viscount Samuel said that they would have to safeguard against the danger of industrial development becoming an eyesore in the Highlands, as he anticipated that there would be an immense increase in international air travel after the war and tourism could become an important industry. This was the recurring dilemma for those who hoped electricity would revitalize the Highlands without, at the same time, taking away their rugged beauty.

The Board was created in 1943 in the midst of war and could do little until the return of peace. When the electricity industry was nationalized the Board remained independent of the British Electricity Authority. In time the Grid reached into the Northern Area and from the 1950s onwards the 'Hydro', as the Board and its works became known in Scotland, constructed and ran more than fifty new water-powered stations. It was, in retrospect, a bold experiment in which the potential of the new technology of electricity was used in an attempt to transform the lives of people living in a remote region. Tourism, as Viscount Samuel had predicted, became an important industry in the Highlands, while the electro-chemical industries went into decline in the face of foreign competition. It turned out that the hydro schemes were vital for the modernization of hotels and other holiday accommodation, and that the building of dams and the harnessing of falls and rivers was not the environmental disaster some had envisaged.[11]

A huge amount of public money was invested in the Highland hydro schemes, and as an experiment in bringing power to a relatively small number of people it was, perhaps, extravagant. However, the technology of electricity generation continues to evolve rapidly, giving rise to new political issues. Since concerns about global warming have been raised, hydro-electric power is now classified as one of the 'green' sources of energy: it might sully the landscape of Scotland but at the same time help save the planet from catastrophic and rapid climate change. In fact the Highland wilderness is now regarded as a potential source of other forms of 'green energy', most prominently the recently erected wind farms that now supply more power to the Grid than hydro-electric stations. There are also plans for extensive wave power stations on the west coast, which is battered by Atlantic gales. However, everything comes at a price. The upgrading of a grid between Loch Ness and the Firth of Forth has given rise to a huge environmental battle over the giant pylons that will stride across the Cairngorm mountains.

CHAPTER 18

THE PROMISE OF CALDER HALL

On 8 October 1956, the *Daily Express* headlined on the front page an exclusive written by its star reporter, Chapman Pincher: 'A-POWER: IT'S HERE! IN ACTION!' More prosaically the sub-heads read: '*The first dinners are cooked – BRITAIN WINS THE RACE.*' The Americans and the Russians had produced electricity with nuclear power experimentally, but Britain was the first to have commercial atomic power with the opening of the Calder Hall station in Cumberland.

'The men using the Calder Hall canteens are eating food cooked with atomic electricity... It is an achievement comparable with the invention of the locomotive – and with possibilities as great,' wrote Pincher. 'The giant uranium furnace which has been producing high-pressure steam for more than a month was secretly linked with the dynamos last week. Top men of the project stood by as steam roared through the turbines, and the dynamos began to hum. As the output mounted, more lamps, more heaters, and more machines were fed with the new power on which so much of Britain's industrial future is being staked.'

The national newspapers were nearly all wildly enthusiastic about the prospects of nuclear power, despite the fact that Calder Hall had been built to make bombs. It was barely a decade after the atomic destruction of Hiroshima and Nagasaki, which brought the war with Japan to an end. The left-wing *News Chronicle* ran a piece by the president-elect of the British Association, Professor P. M. S. Blackett, with the headline: 'More power for Britain – in the nick of time.' The belief that not only Britain but the rest of the world was in urgent need of a new source of power to drive its generating stations was widespread. 'It is nearly true to say that the prosperity of any nation is proportional to the energy at its disposal,' wrote Professor Blackett. 'Vastly more energy will be needed to maintain and advance the prosperity in the next few decades than can easily be got from coal. For the world as whole, nuclear power has come at the right time: for Britain only just in time.'[1]

Quite oblivious, it seemed, to the potential dangers of nuclear power or its economic costs, the professor had a vision of the future: 'The world in which our children and grandchildren will live depends on our efforts now and on the legacy of material power and scientific and technical know-how which we bequeath to them. The houses they will live in, the clothes they will wear, the health they will enjoy, the leisure in which they will be able to cultivate and appreciate the worthwhile and beautiful things of life... all these things will depend on our material command over nature and especially on the amount of energy we can extract from the natural world.'[2]

This was the promise of Calder Hall. But its use for generating electricity was secondary, almost an afterthought. Its origins can be traced back to the early part of the Second World War, when the possibility of releasing a huge amount of energy locked away in the atomic structure of radioactive substances had first been suggested by an elite group of scientists working in England. Several of them were refugees from

Nazi Germany. They had come to the conclusion that it would be possible to control and contain a nuclear chain reaction in such a way that it could be used to make a 'super-bomb'. Work had started in earnest in 1941 to examine the practicalities of manufacturing and testing such a terrifying weapon.[3]

The secretive group became known as the MAUD Committee, which for a long time afterwards was thought to be some kind of acronym. In fact it arose because of the misinterpretation of a telegram sent by the Danish physicist Niels Bohr to another atomic scientist, Otto Frisch, asking that his message be passed on to 'Cockcroft and Maud Ray Kent'. John Cockcroft was an eminent British atomic scientist, but nobody had heard of Maud. As Bohr was still in German-occupied Denmark in 1941, it was assumed that 'Maud' was code. In fact it was the name of Bohr's children's nanny, who had made it to England. But the name stuck.

The MAUD Committee was concerned chiefly with the urgency of making the bomb. It was theoretically feasible, but in the midst of war virtually impossible. In America President Roosevelt had been alerted to the danger of Germany producing an atomic bomb by a letter from Albert Einstein and the Hungarian physicist Leó Szilárd, but there appeared to be no urgency for America to develop a nuclear weapon in the summer of 1941. In fact when the MAUD Committee report was sent to America it was filed away and not given any serious consideration until a British scientist crossed the Atlantic to see why it had been ignored. At the time the United States was still a neutral nation. Then, on 7 December 1941, the Japanese attempted to wipe out the American Navy stationed at Pearl Harbor in Hawaii. The next day the United States declared war on Japan and on 11 December Germany and Italy declared war on the United States.

From 1942 a huge programme was developed to produce an atomic bomb. It could only be made in the United States, as Britain did not have the resources or the huge wilderness sites in which tests could take place.[4] Codenamed 'Manhattan', this project produced the bombs that destroyed Hiroshima and Nagasaki in August 1945. Although much of the expertise that had gone into the making of these nuclear weapons was British, at the end of the war there was no British bomb. This was regarded by the newly elected Labour government, led by Clement Attlee, as dangerous. Soviet Russia, as well as the United States, would have nuclear weapons soon, and Britain would be left at the mercy of these two super-powers. And so began the secret programme to develop the independent nuclear deterrent.

The basic material for an atomic bomb is uranium, which occurs naturally and is mined. In its raw state it cannot be turned into a potentially unstable and explosive material. It has to be refined or 'enriched' to produce either plutonium or uranium 235. This is brought about in a nuclear reactor, which breaks down the uranium into a number of different elements. To build an independent nuclear deterrent, Britain had to have the reactors to produce plutonium. The first of these reactors, or 'nuclear piles', as they were known, was built on the wild coast of Cumbria at a place called Sellafield, which was renamed Windscale in case of confusion with another nuclear site in Lancashire, called Springfields.

On 3 October 1952 the first British-made atomic bomb was detonated in the hold of a frigate moored 600 kilometres off one of the Monte Bello islands, which lie close to the north-west coast of Australia. It was regarded as a qualified success. A leader in *The Times* applauded the recently defeated Labour Prime Minister, Clem Attlee, and his Cabinet for taking the decision to develop an independent nuclear deterrent. 'Britain has proved that she can make atomic bombs,' the editorial stated. '...The cost to Britain of atomic research and production since 1945 seems to have been "something well over £100 millions" by the Prime Minister's reckoning. Parliament has had virtually no control of this spending, and apparently can expect no more in the future. But it can insist that in view of the large sums involved urgent thought be given to the proper deployment of effort in this field.'[5] Winston Churchill had returned as Prime Minister after the defeat of Labour in 1951, and there was no political dispute about the need for Britain to have its own bomb.

Once Britain had embarked on its programme of producing nuclear weapons, it had to invest in many more nuclear reactors to provide the supply of enriched uranium or plutonium. This raised a host of questions about the types of reactor that would be most appropriate, where they could safely be built, and what they would cost. After two more atomic bomb tests in the Monte Bello islands, the government produced, in 1955, a White Paper on the peaceful use of atomic energy. That this was a possibility was already recognized by the MAUD Committee, as well as in America, where some research had been done. In 1954 the British government had created an Atomic Energy Authority which had as part of its responsibility the task of designing nuclear power stations for the generation of electricity. The first of these was already under construction close to Windscale, on what became known as the 'atomic coast' of Cumbria. It was a remote site set in 245 acres of mostly upland farmland.[6]

Calder Hall was certainly the world's first industrial-scale nuclear power station, and appeared to promise a bright new atomic future for electricity generation in Britain. Describing the ceremonial inauguration of the power station, *The Times* correspondent waxed lyrical: 'Today, with a boisterous wind to display the flags – and nearly wreck the marquees – the colourful and almost Wellsian-looking installation deeply stirs the imagination. Truly it has been described as a "courageous enterprise": for Calder Hall represents the inauguration of a comprehensive programme of atomic power stations which, in time, will provide Britain with an ample supply of electricity without the use of coal and oil. Therein lies its magic.'[7]

Her Majesty the Queen, delivering a very carefully worded speech from the temporary dais that had been put up for the opening, was a little more circumspect. 'As the power begins to flow into the National Grid,' she began, 'all of us here know that we are present at the making of history. For many years now we have been aware that atomic scientists, by a series of brilliant discoveries, have brought us to the threshold of a new age. We have also known that, on that threshold, mankind has reached a point of crisis. Today we are, in a sense, seeing a solution of that crisis as this new power, which has proved itself to be such a terrifying weapon of destruction, is harnessed for the first time for the common good of our community.'[8]

The 'harnessing' of this new power did not involve quite the same level of rocket science as the bombardment of uranium atoms in the reactor. What it required was a means of transferring the heat generated in the reactor to boilers that would produce steam to set conventional turbines spinning, which in turn drove the conventional generators. In this sense Calder Hall was a glorified steam engine, as are all nuclear power stations, though the means of drawing heat from the reactor is not always the same. It depends on the process whereby the nuclear reaction is moderated and controlled, so that it might be 'gas cooled', or 'light water cooled', or 'heavy water cooled', or even 'air cooled'.

At the same time that Calder Hall was feeding electricity into the National Grid, it was producing plutonium for Britain's atomic bombs. It also supplied Windscale with some electric power. The target for Britain to become the world's third nuclear power was 200 nuclear weapons, which would necessitate the building of more reactors. But these could be turned to peaceful use as well, following the lead of Calder Hall, which was described officially as 'dual purpose': at one and the same time it could help with the nation's ironing and supply the material to blow a town the size of Coventry to smithereens. But the fact that it had two functions, and that the first was to supply plutonium for bombs, muddied the issue of whether or not it provided an economic means of generating electricity. In fact the sale of power to the National Grid was used to subsidize the cost of the weapons programme, so every electricity bill contained a small contribution to the proliferation of atomic weapons.

The wisdom of spending a fortune on an independent deterrent was questioned at the time of the first Monte Bello test. A *Times* leader on 24 October 1952 commented:

> It is extraordinarily difficult at present to divide expenditure on atomic development into military and non-military compartments. As the Controller of Atomic Energy, Lieutenant-General Sir Frederick Morgan pointed out in March, a great deal of the money spent so far – perhaps 80 or 90 per cent – is common to both... The vast sums of money which have to be spent in the next ten years should none the less be related, with Parliament's consent, to separate military, industrial and medical programmes. Most people will feel, on reflection, that this country will be wise to concentrate from now onwards on the second and third at the expense of the first.

The Times hoped for a reconciliation with the United States so that the Atlantic Alliance would take charge of the weapons programme while Britain could concentrate on the peaceful use of nuclear power. This happened in 1958, but it did not follow that an atomic power station, built primarily for generating electricity, would not produce plutonium for bombs. In reality all nuclear power stations, by their nature, are makers of potentially destructive material. This was one drawback of nuclear power, which otherwise seemed to be *the* solution to Britain's future supplies of electricity. But it was soon clear that there were other difficulties which cast doubt on the benefits of what newspapers in the 1950s liked to call 'atomic electricity'.

From the earliest days of the introduction of electric power to Britain, the quest was to find a way to bring down the price paid by householders or industry. In the home a fierce battle was fought with gas, which provided a well-established alternative for lighting, heating and cooking. Electricity was a luxury until the creation of the National Grid.

Establishing that cheap supply of power to factories and homes entailed the notorious 'march of the pylons' and the building of huge coal-fired power stations in the middle of London and other cities.

More and much larger pylons, put up to carry the post-Second World War Super-Grid, meant that power stations could be built out of town and on top of the coalfields. When electricity at last became really cheap in the 1950s, there was not enough generated and rationing had to be brought in. This was a temporary setback but a warning for the future. New power stations would have to be built, and the prospect of an entirely new source of power that would enable coal to be preserved was all too inviting. There was the alternative of oil-fired power stations, which would be much cheaper to build than nuclear reactors, but the fear of reliance on a fuel that came from the politically volatile Middle East was ever present.

In the mid-1950s about two-thirds of Europe's oil came through the Suez Canal, and when the Egyptian President Abdul Nasser nationalized the canal in 1957 the Conservative government panicked. A disastrous plan to regain control of the canal with Israeli and French forces, but without American backing, ended the career of the Prime Minister, Sir Anthony Eden, and convinced his successor, Harold Macmillan, that Britain should not rely on oil for its power stations. Instead the nuclear programme would be stepped up. The cost of atomic power was not the chief concern at that time. Not only were the nuclear reactors necessary for the production of plutonium, they promised a degree of self-sufficiency in power. Though uranium had to be imported, the sources of supply were not as uncertain as those for oil.

It is not surprising, therefore, that post-war governments were anxious to promote nuclear power and to present the cost of the electricity it produced in the most favourable light. Though the cost of building a nuclear reactor was much greater than that of a coal- or oil-fired station, the running costs were lower. Not much uranium was needed to generate huge amounts of electricity. In theory, a single ton of it could release as much energy as 3 million tons of coal. The fact that nothing like that was achieved in the early nuclear stations built on the Calder Hall model could be put down to the fact that it was a new technology and performance was bound to improve. But the true cost of 'atomic electricity' was not calculated, and the power stations continued to be developed and run by the UK Atomic Energy Authority rather than the Central Electricity Generating Board.[9]

Nuclear power stations were not only expensive to build, there was a potential safety hazard on a quite different scale from that of fossil fuel generators. And then there was the problem of 'back end' costs: what to do with radioactive nuclear waste. It was not possible simply to shut down a nuclear power station, demolish it and grass it over. The by-products of fission were highly toxic and had to be treated or buried, at a cost that would have to be taken into account in assessing the cost of an atomic station.

Because an accident in a nuclear power station was such a hazard, it was not easy to find suitably remote sites for them. Had there been no National Grid they would have been like stranded whales, because there could be no question of placing them in the middle of a city, as Battersea and other 1930s power stations had been. When the Americans were building their first nuclear reactors they ruled that they should be located fifty miles from any town of 80,000 inhabitants, twenty-five miles from one of 10,000 and five miles from one of 1,000.

When Calder Hall was officially opened by the Queen, children from the local schools were given the day off to wave their royalist flags and to see Her Majesty in the flesh. One year later they were being told that they could not drink the milk from the local farms because it had become contaminated with radioactive iodine. On 14 October 1957 *The Times* ran the headline: 'MILK FROM FARMS NEAR WINDSCALE STOPPED RADIO IODINE CONTENT SIX TIMES PERMISSIBLE LEVEL.' One of the reactors had caught fire, and it was discovered that a low level of radioactivity had contaminated a wide area of Cumbria.

After a series of safety checks, the ban on milk sales to children was extended to a region of 200 square miles. Though this appeared to be an isolated instance of contamination, subsequent histories have revealed that Windscale was an accident waiting to happen. The technology was not well understood, and the engineers and scientists operating the reactors had to deal with a variety of unanticipated problems. 'Cores' in the reactor had overheated on a number of occasions and a routine had been established to deal with the problem. But it never occurred in quite the same way twice. The fire occurred unexpectedly during one of these routine cooling exercises.[10]

From time to time, checks on radioactivity around the power station had picked up low levels of contamination which were not thought to be harmful. Even after the fire and the milk ban, the leak of radioactivity was not considered to be very serious. Dr W. G. Marley, head of the health physics division at the atomic research station at Harwell, was quoted in *The Times* as saying that the radioactivity resulting from the leak was considerably less than the 'background level' in many other parts of the world – particularly India and Brazil. The chief atomic safety officer, Mr F. R. Farmer, reported that only two people living in the area had taken advantage of the offer to be medically examined. Only a few employees at Windscale were contaminated, said Mr Farmer, and in most cases a wash with soap and water was enough to give them a clean bill of health.[11]

In retrospect, the reaction to Windscale was remarkably unflustered. The accident and the contamination of milk did not lead to any reappraisal of the bold claim of Rab Butler, Lord Privy Seal, at the opening of Calder Hall that by 1965 every new power station was likely to be atomic. This, many predicted, was the start of a new atomic age, taking over from the coal age of the Victorians. A public information film released not long after the opening of Calder Hall, with the title *Atomic Achievement*, waxed lyrical on this theme: 'The foundation of Britain's Industrial Revolution was coal, and for over 100 years the demand for it rose relentlessly as more and more power was needed for an expanding industry. Above all, coal was consumed by the power stations faster and faster. A new source of power was urgently needed and, to meet the challenge, British science and technology turned to the fundamental power of the universe itself; atomic energy – and the new giant went to work in silence.'[12]

The first reactor at Calder Hall had been built in record time: just three years. More were planned, and the atomic millennium seemed not too far away in the early 1960s. The nationalized electricity industry at least had the prospect of breaking its historic reliance on coal.

CHAPTER 19

A STANDARD OF LIVING

In the early 1960s a series of display advertisements appeared in the daily newspapers, boasting of the great progress that had been made since the war in the provision of electricity for the mass of the people. They were the work of the Electricity Council set up in 1957 to oversee the nationalized industry, and they had a distinctly Orwellian tone about them: a state industry blowing its own trumpet.

In one widely published advertisement there was a photograph of an electricity meter with a finger pointing at the kilowatt/hour measure below the dials. 'This gadget measures our standard of living,' the caption read, continuing in patronizing propaganda prose: 'That's right – it's just an electricity meter. But there's no better yardstick of prosperity than the amount of electricity a nation uses. In Britain we *double* our consumption of electricity every 10 years. By 1968 we shall be using four times the electricity we used in 1948. Look at the countries with the highest standards of living – countries like Britain, Canada, Sweden, the U.S.A. They also have the greatest output of electricity per head of population.'[1]

In another advertisement a disdainful-looking lady switching on an ornamental table lamp is captioned as saying: 'Electricity? Same as always isn't it?' The condescending copy answers: 'Not quite. There have been some big changes and they're part and parcel of our rising standards of living. For one thing, we use well over twice the amount of electricity we did only twelve years ago. To achieve this, 62 new power stations have been put into operation since 1948. We have all heard about the new nuclear power stations, but don't overlook the improvement in the conventional coal-burning type. Today we get 18.5 per cent more electricity from a ton of coal than we did twelve years ago.'[2]

The 1947 Act that brought about the nationalization of electricity in 1948 embodied a commitment to bring this new source of power to every corner of the nation and to do so at a reasonable cost. It took twenty years and more for this to be accomplished, but the principle that everyone *ought* to have a supply of electricity was an important aspect of nationalization. Private, profit-making electricity suppliers would never have sought customers in isolated communities. And before nationalization the local authorities that ran their own electricity service were often reluctant to tax their town customers to subsidize rural electrification. In fact any country that had an ambition to provide all its citizens with electricity in the end had to call on public funds to pay for the connections to isolated communities, whose contribution to income would never cover costs. In the United States, as in Britain, government intervention and investment was needed to bring electricity to the nation's scattered farmsteads.

Nationalization of the electricity supply brought about the wiring-up of rural Britain, which made the electrification of the home almost universal by the 1970s. At the same time, an assumption was made not only that everyone wanted electricity in their home but that, once they were wired in, they would use more and more current as the range of electric goods at affordable prices rapidly expanded. After the embarrassment of the 1950s, when peak loads threatened to crash the system recovering from wartime restraints, the sixties were a time when the nationalized industry could boast of the number of power stations it had built.

By the late 1960s about half of all homes had acquired a refrigerator and an electric immersion heater for their hot water. Electric cookers had become popular too. The electric iron was everywhere, and portable electric fires continued to produce a peak-time headache for the electricity suppliers. And by the mid-1960s the great majority of homes had a black and white television set.[3]

When radio had first become popular in the 1920s there had been no need to be connected to mains electricity to tune in: rechargeable batteries provided sufficient power for a valve set, and a crystal set listened to with headphones needed no current at all. With television it was different: if you were not on the mains only a small generator would provide enough power to switch on. Exclusion from the national mania for television viewing proved to be even more frustrating for people in rural areas than the long wait for electric lighting. In May 1961 a *Times* correspondent reported from Norfolk on the 'lure of TV' in the countryside:

> In these days when advertisements urge people to get up to date with all manner of labour-saving appliances, it might be thought that people in remoter parts of the countryside would be longing for the arrival of electricity. So they are, but not principally for the expected reasons of ease and comfort. Television is what they crave. But for the appeal of this magic box the local electricity authorities would find it much harder to persuade folk to pay for the privilege of going on the mains.[4]

Impatient to tune in, Mr G. Melton, a farmer who lived near the village of Outwell on the Norfolk–Cambridgeshire border, had got himself a diesel engine to run a generator for his own private electricity supply for TV and lighting. He was quoted as saying: 'Of course it's expensive. I use about two gallons a night, and I only run it when we have to. It's a great nuisance having to shut it down last thing at night and then go back up the garden in the dark and go to bed by candlelight. But we like having the telly.' Typically this farmer had three electric 'poles' on his land, for which he was paid nine shillings and eleven pence a year by the local electricity board, yet he would be asked to pay a connection charge of £139 when mains electricity finally came to Outwell. In many parts of the country, the local area boards encouraged farmers to take up electricity with line rental schemes similar to the assisted wiring schemes of the 1930s. Rather than pay a lump sum for a connection there would be a small addition to their electricity bills for a few years.

According to this *Times* report, farm workers were leaving isolated cottages and moving into villages that had finally been connected to the Grid by the local area electricity authorities. Their wives could then have all kinds of gadgetry: electric irons, washing machines and a vacuum cleaner, which made housekeeping a good deal less time-consuming. And then there was Mrs Ada Buck, who had lived near Outwell for eighteen years and had had 'more than enough looking after three sons – doing for the laundry in a copper, heating water for baths in a bowl ...and coping with dirty, smoky oil lamps'. Mrs Buck had won an electric iron in the local whist drive and hoped it would not be too long before she could plug it in, along with a television set.

The widespread claim that electricity made running a home much easier than it had been in the past was disputed by feminist writers, among them Ann Oakley, who published a book, *The Sociology of Housework*, in 1974. She found that a sample of women she interviewed about their domestic lives worked an average of seventy-seven hours a week. It was 'one of the great myths of our time', she said, that modern gadgetry had in any way liberated women by freeing them from domestic drudgery. She found in her research that whether or not a housewife had modern appliances made no difference to her working week. At the same time the housewife with the vacuum cleaner complained of the monotony of her labour and the isolation in which she lived with her young children 'and the cat'.[5]

Whether or not women in the 1960s and 1970s were deluded into believing that electrically powered labour-saving devices made their lives easier, the fact is that the demand for washing machines and vacuum cleaners and irons was pretty much universal. If, as the feminist literature argued, the end result was not less work but a much cleaner, brighter household, then it is at least true that electricity led to a rise in the standard of living, as its propaganda claimed. What was pure fantasy in the propaganda, a favourite theme of the Electrical Association for Women until its demise in 1986, was the claim that vacuum cleaners and washing machines were substitutes for 'servants'. In the case of the wealthy this had some truth: as Gertrude Ferranti found, she could get by with fewer servants once she had an electrified home. For those who had never had servants, a vacuum cleaner was much more efficient than a brush or some mechanical sweeper, but the electric cleaner became, if anything, a taskmaster rather than a servant.

This did not, however, inhibit the demand for electrical equipment. By the 1970s shopping and cooking habits were transformed by the appearance in the majority of homes of a refrigerator. It was now possible to keep ice cream and a range of frozen foods in the ice compartment and to buy food that might last a week longer than if kept in a primitive cooler or pantry. The Walls ice cream tricycles with their 'STOP ME AND BUY ONE' slogan disappeared as corner shops were able to stock packaged ice lollies and creams, though the vans with their jingles survived.

It had been a long time coming, but in the 1960s and 1970s the spread of electricity and the acquisition by nearly all homes of some basic 'labour-saving' equipment was very rapid. Gadgets that had been around for many years became commonplace as manufacturing industry began to offer a wide range of models and electricity prices were held at affordable levels, something the Electricity Council liked to emphasize in its advertisements: 'Since 1948 most prices have risen by two-thirds, most pay packets have doubled. But greater efficiency and increased consumption have ensured that the cost of electricity has not risen on anything like this scale.'[6]

This presented the nationalized supply industry with a tremendous challenge, as the demand for power from domestic consumers seemed set to rise so fast that it would be difficult to build enough power stations to keep up. Forecasts for future coal production suggested that if that was the main fuel available then there would be a crisis by the 1970s. There was a reluctance to switch to oil-fired stations because importing fuel could be an expensive commitment when prices were uncertain. The most promising development was nuclear power, and the industry was proud to announce the opening of new generating stations built along the same lines as Calder Hall. The eleven so-called 'Magnox' power stations (named after a common component material) appeared to function well enough, though there was trouble with corrosion of some of the parts.

In the 1970s the assumption of the CEGB was still that the bulk of the country's electricity would come from a new generation of nuclear power stations with some support from coal and oil. There was a small, but useful, additional source of power from across the Channel. As early as 1949 talks had begun between the British Electricity Authority and France's nationalized EDF (Électricité de France) with a view to trading electricity, something that was already established on the continent of Europe. Just after the war the French Grid was already linked with Switzerland, Italy, Belgium, West Germany and Spain. Underwater cables had been used successfully to supply the Isle of Wight with power from the mainland, but the exchange of electricity between countries with different voltages presented a new challenge.

A relatively new technology was used to create the cross-Channel link. Until the 1930s AC (alternating current) was always favoured for long-distance transmission of electricity, as it was easily stepped up by transformers and stepped down for delivery to factories and homes. This only became possible with DC (direct current) with the development in the 1930s of a valve that could transform high currents for sending into lower currents for practical use. It was found, for technical reasons, that high-voltage direct current (HVDC) was more suitable than AC for underwater cables as well as for linking otherwise incompatible systems, and this was eventually the choice for the cross-Channel link as it was developed in the late 1950s.[7]

The link did not come cheaply, the cable and the two control stations, at Lydd on the Sussex coast and Echingen near Boulogne in France, costing around £6 million (over £80 million at today's prices). But both countries felt it was worth it, as by switching the flow of the power between them they could do without an extra power station to meet peak loads. Because of different time zones as well as domestic habits, the peaks were at different times. For example, the midwinter peak demand in France was 8.30 a.m., whereas in England it was 5 p.m. And at the time the link was being built, the peak of supply in the two countries was different. France then got nearly half its supply from hydro-electric stations in the Alps and these were most productive in the autumn and spring, in contrast with England, where the peak for surplus energy was in the summer when alternative demands on coal were at their lowest. In theory the cost of the cable link would pay for itself after a few years, as annual savings were reckoned at £340,000.

Two cables had to be laid across the Channel, kept a critical distance apart. To lay theirs the British used a collier ship named the *Dame Caroline Haslett* in honour of the recently retired director of the Electrical Association for Women. Cheerily the Channel Cable executive committee announced that 'the egg with the lion on it may be boiled by melting Alpine snows, and the Paris Metro could at times be energized by coal from British collieries'.[8] The cross-Channel link was ready for full operation early in December 1961. About six weeks later it was snapped in two by a ship's anchor. It took the *Dame Caroline Haslett* weeks to find the loose end of the British cable and the link was out of action for a long time.

Although it was frequently out of action because of fouling of the cables on the seabed, the cross-Channel link at least demonstrated that in future the sharing of load between France and England was technically feasible and could be of great benefit to both countries. This first scheme kept going until 1982, when it was decommissioned while a new cable, buried under the seabed and capable of carrying much higher voltages, was put in place and began operating in 1985. By that time France no longer relied so heavily on the hydro schemes of the Alps for electricity, and had embarked on the most extensive programme of nuclear power station building in Europe.

Britain might have followed suit, but, for a variety of reasons, the atomic power programme ran into trouble. Quite apart from the question of the true cost of the electricity generated by nuclear fission, and the inherent dangers of contamination and disposal of spent fuel, there was the highly charged political issue of the future of the coal industry. What would happen to the miners if the electricity industry no longer needed their coal? Wiping out what had been a huge industry, employing 1.5 million in 1939, was politically unacceptable. In 1961, the Conservative Prime Minister Harold Macmillan shrewdly offered the chairmanship of the National Coal Board to Alfred Robens, a Labour Party stalwart and former trade union official who had been MP for the mining constituency of Blyth since 1945. There was no doubt that coal mining was to be run down: the question was how quickly and at what social cost. Elevated to the peerage as Baron Robens of Woldingham when he took over as Coal Board chairman, he became an instantly recognizable figure with his trim moustache and dapper appearance.

Brought up in a working-class family in Manchester, Alfred Robens was a self-made man who relished the privileges of power and influence.

He liked to be chauffeured in a Daimler with the number plate NCB1 and had a private plane to take him round the country. He set himself up in Eaton Square, where he luxuriated in the title of Old King Coal. Although he could not prevent the closure of inefficient mines and the shrinking of the industry during his ten-year reign as NCB chairman, Robens did fight hard and with some success to convince the government and the electricity industry that coal-fired power stations were still needed. He argued vociferously that the nuclear power stations were benefiting from false accounting, and made sure coal was offered at competitive prices. His battle with the CEGB came to a head in 1968, when they proposed to build a nuclear power station near Hartlepool, in the heart of coal-mining country. This incensed Robens, who made his disagreement with the government and the Electricity Board public in a series of the letters to *The Times*. In one he offered coal at a price that would produce electricity at a cost lower than that estimated for the most efficient nuclear power station.[9]

When Robens and the miners lost the battle of Hartlepool, it looked in the early 1970s as if nuclear power would begin to replace both oil- and coal-fired generating stations. But the programme to build new, and supposedly more efficient, nuclear power stations ran into technical troubles, while concern about its safety brought about further delays. While expansion of nuclear power was stalled, the coal-fired power stations kept burning, producing 75 per cent of the country's electricity, with the CEGB remaining the single largest customer of the National Coal Board. The dependence of one on the other was brought home miserably in the first weeks of 1974.

A work to rule by miners from the middle of 1973, in protest against a fall in their real earnings, reduced the stocks of coal at power stations and drove up the price of coal. The Yom Kippur war between Israel and Egypt raised the price of coal further when the Arab oil-producing countries put an embargo on sales to those who supported the arming of Israel. By Christmas 1973 there was a danger that the electricity supply industry would not be able to cope with both industrial and domestic demand. In response, the Prime Minister, Edward Heath, brought in an emergency measure that became known as the 'three-day week'. Beginning on New Year's Eve 1973, industry and many commercial companies were ordered to restrict their working hours to conserve electricity and to avoid a complete blackout.

From January to March 1974 the rationing continued, while Heath fought it out with miners. In February an election was called with the Conservative battle cry of: 'Who governs Britain?' Nobody in particular, as it turned out. Neither Labour nor the Conservatives had an overall majority, though Heath was out and Labour's leader, Harold Wilson, became Prime Minister. In a final act of defiance the miners went on strike and won a substantial pay rise, which inevitably raised the cost of coal and, in turn, of electricity.[10]

The rationing of electricity in the 'three-day week' was a temporary setback for the supply industry. When the dust had settled and Labour properly returned to power at the October election of 1974, the downturn in the economy soon made nonsense of the CEGB forecasts for future demand for electricity. It was no longer 'doubling every ten years'. But that is more or less what had been planned for. New power stations were being completed, and as they came into operation the CEGB found it could generate far more electricity than was needed. As an economy measure the Board began to close down the older and less efficient coal-fired stations.

By the late 1970s the dreams of the electricity pioneers had been realized, not the utopian 'all-electric' world of Ferranti's imagination, but a nation wired into its furthest corners and supplied with power that was no longer an expensive luxury. That might have been the end of the story of the electrification of Britain. But just at the point in history when electricity became commonplace, it was the subject of a host of political upheavals far greater than those which had greeted its arrival in late Victorian Britain.

CHAPTER 20

BACK TO THE FUTURE

In 1985, as Mrs Thatcher's programme of privatizing much of British industry was in full swing, the grand old man of Tory politics, Harold Macmillan, made a characteristically witty speech in which he likened the sell-off of state assets to an impoverished aristocratic family meeting its debts by disposing of its precious belongings. 'First of all the Georgian silver goes. And then all that nice furniture that used to be in the salon. Then the Canalettos go.'[1] This speech, at the Royal Overseas Club, was widely reported as a criticism of privatization. Macmillan, who was then nearly ninety and ennobled as Lord Stockton, protested that he had been misunderstood. He was all in favour of returning nationalized industries to the cut and thrust of private enterprise. What he thought was inadvisable was that Mrs Thatcher's government appeared to be treating the billions the sale of state assets put into the Treasury coffers as income rather than capital: they were spending it rather than investing it for the future. Macmillan's back-track had little effect: 'selling the family silver' struck a chord which, in retrospect, was not entirely inappropriate.

There was plenty of family silver to dispose of – British Telecom, the railways, the mines, water and the gas industry. But the biggest asset of all in the nation's treasure chest was the electricity industry. The Thatcherite view was that the nationalized industry was essentially inefficient, committed to building huge power stations, over-estimating demand and charging too high a price for the current generated. The twelve regional electricity boards had to buy from the CEGB, which had a virtual monopoly on generation. Without competition there was no price mechanism to ensure the production of electricity was as cheap and efficient as possible.

If, as sometimes appeared to be the case, Mrs Thatcher and the enthusiasts for privatization in her Cabinet imagined that they would be reviving 'Victorian values', they were mistaken. As earlier chapters have shown, there was as much enthusiasm in the early days of the industry for public ownership as there was for a competitive private market. In fact, it was difficult to imagine how the structure of the electricity industry Mrs Thatcher's government inherited in 1979 could be privatized without causing some huge upheavals. How could you introduce competition into a system that had to function continuously and co-operatively through the National Grid? Electricity has to be on tap continuously in sufficient quantities to meet the periods of greatest demand. Above all other considerations, there was the fact that any proposal to privatize the industry would be a minefield: literally.

Despite all the hopes for nuclear power as an alternative to coal, at the time Margaret Thatcher took office it produced only about 20 per cent of the nation's electricity. The National Coal Board, still in state ownership, provided nearly 80 per cent of the CEGB's fuel. This did not change when the Conservatives won the 1979 election. Mrs Thatcher supported the British coal industry, the government investing heavily in the modernization of the pits. The only complaint she and her Cabinet voiced was that too many uneconomic pits were being kept open for social reasons and this, in turn, pushed up the price of electricity. But it was not a major issue during Mrs Thatcher's first term of office. Her political position was tenuous, since opinion polls indicated that she was hugely unpopular. Victory in the Falklands war changed that, and she was returned with a landslide majority in 1983.

The confrontation with the National Union of Mineworkers in 1984, which gave rise to some of the most violent and shocking incidents in post-war British history, was fundamentally about destroying the hold miners had on the electricity industry. There would be no point in privatizing electricity while it was dependent on home-produced coal for the bulk of its supply. Edward Heath's 'three-day week' and defeat at the hands of the miners in 1974 were still fresh in the memory.[2]

Mrs Thatcher did not rush into battle. Anticipating a showdown at some stage, the government had made sure that, by 1984, there were good stocks of coal at the power stations, enough for about four or five months. In addition, three changes to the law would make it more difficult for the miners to take strike action. The 1982 Employment Act outlawed the picketing of places or firms other than those that employed strikers directly, which was generally known as 'secondary picketing'. In effect miners were breaking the law if they picketed power stations in an attempt to get the workforce to back the strike or to stop coal being delivered. The same Act overturned a 1906 law that had given trade unions immunity from claims for damages resulting from strike action. Union funds could be sequestered by the courts if they refused to pay up. A further Trade Union Act in 1984 made it illegal for a trade union to call a strike without first balloting its members.

A tough negotiator, Ian MacGregor, a Scot who had spent much of his career in America, was put in charge of the National Coal Board with the task of turning the heavily subsidized industry into a concern that might at least break even. This would involve closing a large number of pits, twenty in the first phase, throwing 30,000 miners out of work. Yorkshire miners went on strike, and the industrial action spread rapidly. The president of the National Union of Mineworkers, Arthur Scargill, called a national strike, but without holding a ballot. The mineworkers' union was a regional federation with different rules and traditions in different parts of the country. Some miners, notably those in Nottinghamshire, which had some of the most profitable pits, refused Scargill's call to down tools. There were violent clashes between the strikers and the miners who kept the pits going.

The strike lasted for nearly a year, drawing to a close in March 1985 when there had been a drift back to work. To have succeeded the miners would have had to close down the power stations. In desperation they appealed to the public to crash the system by turning on everything they could at times of peak demand. The fact that this did not happen adds credence to the belief that the government had planned for a major fight. It is remarkable, given the dependence on coal at the time of the strike, that electric power never failed. A post-mortem revealed how that was achieved. Some power stations were adapted to burn fuel oil, and this was estimated to have saved 38 million tons of coal. Stockpiling provided 13 million tons, and imports of coal another 10 million tons. But the biggest single contribution was from the pits that kept working. They produced 42 million tons of coal during the strike year.[3]

The rapid closure of pits that followed the defeat of the NUM in 1985 did not put an end to the electricity industry's reliance on coal at the power stations. While mining communities suffered great hardships, coal, most of it home-produced, continued to keep the lights on and industry and transport running normally. Oil and gas were insignificant fuels in 1985, and it was only the nuclear power stations that generated substantial amounts of electricity.

This situation did not change radically until the government finally got round to selling off the electricity supply industry. Ironically, it was the National Grid, a creation of the state, that made a market in electricity possible. The power produced by the generation companies was 'pooled', as it had to be if the national system of distribution were to be continued. The National Grid became a private company owned, at first, by the Regional Electricity Companies (RECs), which were, in effect, the privatized former area boards of the state system. In place of the CEGB were two large private generation companies: National Power, which took just over half the generating capacity, and PowerGen, which took around a third. These were the main suppliers to the Grid. Scottish Power and EDF in France were minor suppliers, as were the RECs, which were given the option of generating some power for themselves. Nuclear power, when exposed to commercial scrutiny, proved impossible to privatize at first and remained in public ownership for a number of years under the title of Nuclear Electric.

The privatized companies were sold to the public, who took up the shares enthusiastically and joined the 'shareholding' democracy that the Thatcherites were keen to promote. Within five years the value of these shares rose by over 250 per cent, out-performing the stock market by 100 per cent. They were clearly a bargain, and provided a few thousand individuals with a handsome profit. In the same period the electricity supply industry was transformed.[4]

In 1990, 92 per cent of the fossil fuel used for electricity generation was coal, 7 per cent was oil and 1 per cent was gas. Over the next five years the private companies' orders for British coal fell from 74 to 30 million tons, and coal prices fell by 20 per cent. At the time of the 1984–5 strike there were just under 250,000 miners in Britain. In 1994 the mines were privatized and the companies that succeeded the National Coal Board employed only 7,000 miners. Parliament recognized that in social terms there was a 'coal crisis', but the government was determined to stick with the harsh judgement of the market. Exception was made only for nuclear power, which was protected.

It was an assumption of Mrs Thatcher and her government that by the time the coal industry had been reduced to a commercially viable rump, nuclear power would be available to take up the bulk of electricity generation. But it was not to be.

The first Magnox nuclear power stations, built along the lines of Calder Hall, had been a modest success but were reaching the end of their life. Magnox reactors had been fuelled with raw uranium, which was enriched to provide the material for atomic weapons. New, and potentially more efficient, designs of power station were pioneered in the 1960s, using enriched uranium as fuel. There was a variety of models available, two developed in the United States, known as the BWR (boiling water reactor) and the PWR (pressurized water reactor). Britain, however, decided to go its own way and hoped to produce an alternative that would compete on the international market with its rivals. This was the AGR (advanced gas-cooled reactor), and the first to be commissioned as part of a new and expanding nuclear programme was sited alongside an old Magnox station at Dungeness, a headland of shingle in Kent, on the south-east coast of England.

The contract to build the AGR, modestly named Dungeness B, was awarded in 1965 with a schedule that would have it producing electricity alongside the Magnox station by 1970. But the novel design soon ran into a host of technical problems.[5] In 1969 the consortium formed to build the AGR went bust and the CEGB had to take over. By 1975 the estimated date for completion was set at 1977. The original budget for the power station had been £100 million. By 1977 it was estimated at £280 million and two years later, when it was still not in operation, at £410 million. Dungeness B did not feed any electricity into the Grid until 3 April 1983, by which time it had cost £685 million. Meanwhile the nuclear power programme both in Britain and abroad ran into serious trouble when an American pressurized water reactor, which was becoming the best-selling model worldwide, and was reputedly exceptionally safe, suffered a near calamitous meltdown.

Three Mile Island, Unit 2, PWR nuclear power station began operation at the end of December 1978. Like all such power stations it was sited well away from any large population, on an island in the Susquehanna River, three miles downstream of Middletown, Pennsylvania. It had been in operation for just three months when it was discovered that something had failed in the complex workings of its cooling system and that the uranium core had gone into meltdown. The result was a release of radioactivity that led to the closing of schools in the area and precautions taken with farm livestock. An investigation by the US Nuclear Regulatory Commission concluded that there were no deaths or serious illnesses associated with the fallout. But this reassured neither the public nor the government. Whether or not there was an increase in cancer rates, as some further studies claimed, it was obvious that there could have been a major disaster at Three Mile Island.

The loss of confidence in the safety of nuclear power after this radiation leak brought an abrupt halt to America's ambitious nuclear power programme. In the 1960s the forecast was for 1,000 reactors by the start of the twenty-first century. By 2008 there were only 104 operating, supplying about 20 per cent of United States electricity. The pressurized water reactor was not the commercial marvel it had once been thought to be.

In Britain, however, the Three Mile Island accident caused barely a ripple of concern in government circles. Margaret Thatcher became Prime Minister on 4 May 1979, a few weeks after the American accident had been in the news, and promptly adopted a pro-nuclear policy. The country's first pressurized water reactor on the Three Mile Island model was planned for the site of an existing Magnox reactor adjoining the village of Sizewell, on the coast of Suffolk in East Anglia. Before it could go ahead, however, there was to be a public inquiry at which concerns about the safety of nuclear power, and the PWR design in particular, could be examined.

The Inspector appointed to conduct the inquiry was the distinguished planning lawyer Sir Frank Layfield, who began to take evidence in January 1983.[6] It proved to be the longest planning inquiry in British history, with hearings continuing for more than two years. Layfield took another two years to write up his 3,000-word report, which was published in January 1987 at a final cost of something like £10 million. On 26 April 1986, while Layfield was writing his report, the worst disaster in the history of nuclear power occurred at Chernobyl in the Ukraine, then part of Soviet Russia. One of the reactors exploded, releasing massively more radiation than the Hiroshima bombs and sending a dust cloud high into the atmosphere, where it drifted on the wind over a wide area of northern Europe. In the Ukraine more than 300,000 people were evacuated from the most polluted regions, and weather-borne radiation found its way into Lapland reindeer and lamb in the uplands of Britain.

Chernobyl was not mentioned in Layfield's report because it occurred after the close of the hearings and could not be included in evidence. But it was a serious setback for the nuclear power industry. Layfield, in fact, came down in favour of building the PWR at Sizewell, and calculated that it had a good chance of providing electricity more economically than a coal-fired station since fossil fuel prices were rising. Work on the building of Sizewell B began in 1988 and it was in operation by 1995.

Meanwhile, as the nuclear power programme floundered, the privatized electricity industry found that there was a cheap and instantly available source of fuel for its generators to replace the rapidly diminishing supply of coal. By chance it came from electricity's great rival, the gas industry. In March 1967, British Gas began to publish in the newspapers advertisements that featured a photograph of a frying pan in which two oatmeal-coated herrings sizzled over the flames of an oven hob. The caption read: 'There are two very valuable catches in the North Sea. This is one of them being cooked by the other ...Already enough has been found to enable the gas industry to plan the conversion of its distribution system so that eventually natural gas will be used "neat" throughout Britain... The gas industry is ready to make the most of this new indigenous source of energy – ready for the new age of High Speed Gas.'

North Sea gas was methane with different properties from the town gas extracted from coal that had sustained the industry since the early nineteenth century and was principally hydrogen. Once the government was convinced that there was sufficient natural gas to replace coal gas, a programme to convert millions of cookers and other appliances was begun. Rapidly the old gas works disappeared, along with most of the gasometers that had been such a feature of the Victorian urban landscape. A natural gas Grid was established to distribute supplies around the country direct from the North Sea. (It is now owned and run by the same company that runs the electricity Grid.) There was never any intention that this gas would be used to fuel power stations: on the contrary, in both the United States and in Europe, natural gas was classed as a 'premium fuel', reserved for domestic consumers and industry. But that changed in the 1980s. In America, new reserves of natural gas were found and the restriction on its use was lifted. Once it was available for the generation of electricity, a new, and very efficient, kind of power station was developed. On the same principles as a jet engine, methane gas could be used to power a turbine that could be coupled directly to a generator without steam. However, the heat from the turbine could be used to create steam to drive a secondary generator. These natural gas power stations became known as CCGTs – combined cycle gas turbines.

Restrictions on the use of natural gas for power stations were lifted in Britain in the 1980s, and the inviting prospect of building CCGTs was open to the commercial companies of the newly privatized electricity industry. Supplies of North Sea gas were still abundant, and the gas turbine power stations could be put up very quickly with American technology available more or less 'off the peg'. Rapidly, gas-fired generators began to supply the Grid with a significant proportion of its power. Within the industry this sudden switch from coal to natural gas as a fuel for electricity generation became known as the 'dash for gas', so rapid and widespread was the building of CCGTs.[7]

At the beginning of 2010, the mix of fuels used to produce the electricity that is distributed from power stations through the National Grid is roughly in the following proportions: gas 42 per cent, coal 40 per cent, nuclear 15 per cent, wind 1 per cent, pumped storage hydro 1 per cent and hydro (not pumped storage) 1 per cent. These proportions vary continuously throughout the day. Nuclear's contribution might rise to 18 per cent and coal's fall to 34 per cent. Wind at times will contribute practically nothing, at other times up to 1.5 per cent. Current from Europe is mostly less than 1 per cent, and occasionally there is some from Northern Ireland.

In his memoirs, Cecil Parkinson, who was Secretary of State for Energy in 1988, wrote: 'I have never understood the argument that Britain somehow owes a great debt to the mining industry. The industry was given a privileged position and it abused the privilege.'[8] Parkinson made it clear that the privatization of electricity had as its ultimate aim 'the end of the political power of the National Union of Mineworkers'. And it was quite true that when Britain was utterly reliant on home-produced coal for its electric power, the only threat to supplies was militant action by the miners. They could, as Mrs Thatcher put it, 'hold the country to ransom'.

However, the policy of supporting the miners, which both the Conservative and Labour parties adopted after 1945, was neither entirely sentimental, as Parkinson suggested, nor simply timid, as Mrs Thatcher imagined. The only alternatives to British coal for electricity generation were nuclear power, or imported oil and coal. Each time there was an 'oil crisis' due to conflicts in the Middle East, the case for British coal was strengthened. The price of oil fluctuated wildly and, though it was often cheaper as a power fuel than indigenous coal, there was no telling when the price might shoot up again. Similarly, imports of coal might be cheap, but how reliable were they? For a nation that had had to contend with two world wars in the first half of the twentieth century, the prospect of being dependent for its essential electricity supplies on fuel imported from countries that might at any time turn hostile was an unhappy prospect. British coal miners seemed less threatening than the oil sheikdoms of the Middle East.

The privatization of the electricity industry, and the defeat of the miners, marked the end of this concern about where we get our fuel. At first there was little difficulty in maintaining home-produced supplies because we had our own natural gas reserves. Gas replaced the generating power lost to coal. Now the British reserves are running out and gas is already being piped in from Norway and Europe. The fuel for nuclear power stations is imported, though supplies of uranium are not considered to be a potential problem. But a large number of existing nuclear power stations are being decommissioned and it takes time to build new ones. Coal still makes a vital contribution to the Grid, but something like 70 per cent of the coal used for electricity generation is now imported. Oil is not a fuel anyone wants to rely on, and, in any case, fossil fuels are being phased out to comply with carbon emissions regulations.

It is not possible to say exactly what our dependence is now on imported fuel and power for our electricity supplies, but we certainly could not survive without imported natural gas. In about ten years' time we will be almost entirely reliant on gas supplied from abroad. We could reopen mines in Britain – there are substantial reserves of coal – but the fuel they produced would be both expensive and environmentally unacceptable. Wind farms and other 'renewables', as well as a determined effort to reduce demand for electricity, could ease the problem. But it is hard to envisage a future without a much greater contribution from nuclear power, and inevitably it is that direction that the most recent government policy has taken. The Cabinet paper 'The Road to 2010', published in July 2009, was emphatic: 'Nuclear power is a proven technology which generates low carbon electricity. It is affordable, dependable, safe, and capable of increasing diversity of energy supply... Combating climate change, the single greatest threat to humanity this century, requires a much greater role for low carbon fuels in the global energy supply than before... Nuclear energy is therefore vital to the challenges of sustaining global growth, and tackling poverty.'[9]

Whereas privatization had the effect of diminishing the importance of nuclear power in Britain, the issue of global warming has redeemed it, so that it is now vital to the government's plans for the electricity industry. Though the government insists that there will be no subsidies for the building of new nuclear power stations, the regulation of carbon emissions will ensure that there is no alternative for companies supplying electricity in the future. The byword of privatization – 'deregulation' – did not last long. As the Victorians recognized, it is impossible to treat electricity like a commodity: it is a resource that no government can abandon to the whims of the market.

As long as the lights stay on, and the price of electricity does not rise too steeply, the British public will not be too concerned about how its power is produced. The vociferous environmental lobby will campaign for more 'renewables', though wind farms and all other 'green' forms of electricity generation have their drawbacks too. One or two wealthy individuals may even become latter-day successors to pioneers such as Siemens and Armstrong who generated their own electricity. They will, of course, be much better off than their Victorian forebears: they will have much more sophisticated technology and also the opportunity to sell power back to the Grid. But, in the grander scheme of things, their contribution will be negligible and hardly relevant to the majority of the population – unlike Regulation 244/2009, a directive recently handed down by the European Union and one that affects the heart of every British home.

EPILOGUE

The announcement that the familiar little filament light bulb was to be outlawed by the European Union was greeted with shock and indignation in Britain: this was yet another petty-fogging interference in our everyday lives by the bureaucrats in Brussels. There were reports of the stockpiling of 'traditional' light bulbs. Experience with the alternative, low-energy lamps had often been disappointing: there was a loss of the familiar brightness. But we are having to learn to live with them. Already the more power-hungry 150- and 100-watt bulbs have been taken off the shelves, and no filament lamps will be sold after the end of 2012.

It is fitting that the incandescent lamp, among the clutter of electrical appliances now found in nearly every home, should be singled out. The invention of Swan and Edison, so novel and exciting in Paris in 1881, was the first electrical gadget that required substantial amounts of power. Neither the electric telegraph nor the telephone gave rise to power stations. It was the proliferation of the filament bulb, first in hundreds, then in thousands and then millions, that created the demand for electricity in large quantities. Though electric lighting is just one demand on modern power stations, which supply current to a huge range of domestic appliances as well as to industry and transport, it remains hugely significant.

In the Electricity National Control Centre, run by the privately owned National Grid, the rise and fall in demand for power is monitored continuously twenty-four hours a day. An illuminated diagram the size of a large cinema screen shows the interconnections between all the power stations in the country. Here the pattern of British life is reflected in the rise and fall of demand for electricity. It is the peaks that are most critical: the electricity fed into the Grid has to be sufficient at all times to meet demand. Electricity cannot be stored.

The switching on and off of the nation's lights remains a significant element in demand. It is not just a matter of night and day. We are so accustomed now to a decent quality of light at all times that a dull, overcast day will bring the lights on and prompt a rise in demand for power that the National Grid engineers have to match. The conversion of light bulbs to energy-saving models is not, therefore, mere tinkering. We might not like the light cast by energy-saving bulbs, and rue the demise of the filament lamp, but the necessary adjustment of our vision will be surely no more demanding than that experienced by the Victorians who winced and wondered at the glow of the first electric illuminations the world had ever seen.

Now that our daily lives are entirely reliant on a continuous supply of electricity, it is surprising to learn how unsure earlier generations were about its advantages over gas and coal in the home and for industry and transport. As with so many novel technologies, electricity was not embraced at first with any great enthusiasm, while there was a good deal of anxiety about what might happen if it fell into the hands of monopoly capitalists. Britain was slow to adopt this new source of power, and even in the 1920s a high-handed approach by government was necessary to impose on the country the grid of pylons and cables that ushered in the era of cheap and affordable electricity. It was not until the 1960s that electricity became almost universal, and indeed essential for a modern standard of living that required a range of gadgetry only it could power.

No sooner had the dreams of the electrical messiahs of the 1920s been fulfilled, however, than the problem of sustaining supplies of cheap electricity for all became critical. There is now a wide range of technologies capable of producing electrical power, but there is no system without its drawbacks. This is a frustrating characteristic of electricity: in itself it is smokeless, powerful and brilliantly versatile, yet it can only be generated in any quantity by more troublesome and less versatile technologies. And the biggest drawback of all is that, unlike coal, or gas or oil, it cannot be stored except in very small amounts, which means that it always has to retain a capacity far in excess of that required most of the time.

The demise of the filament bulb is no doubt just the first of the many changes we will have to make in our attitude towards electricity. It is inconceivable that this strangely mercurial source of power will not remain an essential part of our world, but we will almost certainly have to use less of it and find less environmentally damaging ways of producing it, as we strive to retain the astonishing luxuries it has made possible in our daily lives.

BIBLIOGRAPHY

Appleyard, Rollo, *Charles Parsons, His Life and Work*, Constable, 1933.

Arnold, Lorna, *Britain and the H-Bomb*, Palgrave, 2000.

Arnold, Lorna, *Windscale 1957: anatomy of a nuclear accident*, Macmillan, 1992.

Ballin, Harold Hanns, *The Organisation of the Electricity Supply in Great Britain*, Electrical Press, 1946.

Barker, T. C., and Robbins, M., *A History of London Transport: passenger travel and the development of the metropolis, vol. 2: The Twentieth Century to 1970*, Allen & Unwin, 1974.

Bowers, Brian, *R. E. B. Crompton: an account of his electrical work*, HMSO, 1969.

Bowers, Brian, *A History of Electric Light and Power*, Perigrinus, 1982.

Bowers, Brian, *Lengthening the Day*, Oxford University Press, 1998.

Bowers, Brian, *Michael Faraday and Electricity*, Priory Press, 1974.

Bragg, Sir William Lawrence, *Electricity*, G. Bell & Sons, 1947.

Byatt, Ian Charles, *The British Electrical Industry 1875–1914*, Clarendon Press, 1979.

Chandler, Dean, and Lacey, A. Douglas, *The Rise of the Gas Industry in Britain*, British Gas Council, 1949.

Chick, Martin, *Electricity and Energy Policy in Britain, France and the United States since 1945*, Edward Elgar, 2007.

Cochrane, Rob, *Pioneers of Power: the story of the London Electric Supply Corporation*, London Electricity Board, 1987.

Cochrane, Rob, *Power to the People: the story of the National Grid*, Newnes, 1985.

Crompton, Rookes Evelyn, *Reminiscences*, Constable, 1928.

Dillon, Maureen, *Artificial Sunshine*, National Trust, 2002.

English Electric Company, *A Collection of Letters to Sir Charles William Siemens, 1823–1883*, 1953.

Ferranti, G. Z. de, *The Life and Letters of Sebastian de Ferranti* (written with Richard Ince), Williams & Norgate, 1934.

Forty, Adrian, *Objects of Desire: design and society 1750–1980*, Thames & Hudson, 1986.

Friedel, Robert D., Israel, Paul, and Finn, Bernard S., *Edison's Electric Light: biography of an invention*, Rutgers University Press, 1986.

Gooday, Graeme, *Domesticating Electricity: expertise, uncertainty and gender 1880–1914*, Pickering & Chatto, 2008.

Gowing, Margaret, *Britain and Atomic Energy 1939–1945*, Macmillan, 1964.

Hammond, Robert, *The Electric Light in our Homes*, Frederick Warne, 1884.

Hannah, Leslie, *Electricity before Nationalisation*, Macmillan, 1979.

Hannah, Leslie, *Engineers, Managers and Politicians: the first fifteen years of nationalised electricity supply in Britain*, Macmillan, 1982.

Hardyment, Christina, *From the Mangle to the Microwave*, Polity Press, 1988.

Helm, Dieter, *Energy, the State and the Market: British energy policy since 1979*, Oxford University Press, 2003.

Hennessey, R. A. S., *The Electric Revolution*, Oriel Press, 1972.

Hughes, Thomas P., *Networks of Power*, Johns Hopkins University Press, 1983.

Israel, Paul, *Edison: a life of invention*, John Wiley, 1998.

Jackson-Stevens, Eric, *100 Years of British Electric Tramways*, David & Charles, 1985.

Jekyll, Gertrude, *Old West Surrey*, Longmans, 1904.

Klapper, Charles Frederick, *The Golden Age of Tramways*, David & Charles, 1974.

Luckin, Bill, *Questions of Power: electricity and environment in inter-war Britain*, Manchester University Press, 1990.

McDonald, Forrest, *Insull*, University of Chicago Press, 1962.

Nye, D. E., *Electrifying America: social meanings of a new technology, 1880–1940*, MIT Press, 1990.

Oakley, Ann, *The Sociology of Housework*, Martin Robertson, 1974.

O'Dea, William, *The Social History of Lighting*, Routledge & Kegan Paul, 1958.

Parsons, R. H., *Early Days of the Power Station Industry*, Cambridge University Press, 1939.

Passer, H. C., *The Electrical Manufacturers: 1875–1900*, Ayer, 1988.

Payne, Peter L., *The Hydro*, Aberdeen University Press, 1988.

Percival, G. Arncliffe, *The Electric Lamp Industry*, Pitman, 1920.

Pole, W., *The Life of Sir William Siemens*, John Murray, 1888.

Political and Economic Planning, *The British Fuel and Power Industries*, PEP, 1947.

Political and Economic Planning, *The Market for Household Appliances*, PEP, 1945.

Poulter, J. D., *An Early History of Electricity Supply*, Peregrinus, 1986.

Price, Terence, *Political Electricity: what future for nuclear energy?*, Oxford University Press, 1990.

Randell, Wilfrid L., *S. Z. de Ferranti: his influence upon electrical development*, Longmans, 1948.

Roberts, Jane, Elliott, David, and Houghton, Trevor, *Privatising Electricity: the politics of power*, Belhaven, 1991.

Robertson, Alex J., *The Bleak Midwinter: 1947*, Manchester University Press, 1987.

Rowland, John, *Progress in Power: the contribution of Charles Merz and his associates to sixty years of electrical development 1899–1959*, Newman Neame, 1960.

Schivelbusch, Wolfgang, *Disenchanted Night: the industrialisation of light in the nineteenth century*, Berg, 1988.

Scott, John Dick, *Siemens Brothers 1858–1958*, Weidenfeld & Nicolson, 1958.

Self, Sir Henry, *Electricity Supply in Great Britain, its Development and Organisation*, George Allen & Unwin, 1952.

Surrey, J. (ed.), *The British Electricity Experiment: privatisation, the record, the issues, the lessons*, Earthscan, 1996.

Swan, K. R., *Sir Joseph Swan and the Invention of the Incandescent Electric Lamp*, Longmans, 1948.

Swan, M. E., and Swan, K. R., *Sir Joseph Wilson Swan: inventor and scientist*, Ernest Benn, 1929.

Thwing, Leroy Livingston, *Flickering Flames: a history of domestic lighting through the ages*, G. Bell & Sons, 1959.

Wachhorst, Wyn, *Thomas Alva Edison, An American Myth*, MIT Press, 1981.

Williams, Roger, *The Nuclear Power Decisions: British policies 1953–78*, Croom Helm, 1980.

Wilson, J. F., *Ferranti: a history, vol. 1, Building a Family Business 1882–1975*, Carnegie Publishing, 2000.

NOTES

Introduction

1. Eileen Murphy, *The Lure of the Grid. The plain facts about electricity in the home, etc.*, London, 1934.
2. 'Channel Power Cable in Use', *The Times*, 5 October 1961.

Chapter 1: Under the Arc Lights

1. Keith Farnsworth, 'The Illuminating History of a Pioneer Called John Tasker', *Quality* magazine, Sheffield, November/December 1978.
2. *Sheffield and Rotherham Independent*, Tuesday, 15 October 1878. Also the *Sheffield Daily Telegraph* of the same date. Also *Bell's Life in London, and Sporting Chronicle*, 19 October 1878.
3. Michael Brian Schiffer, 'The Electric Lighthouse in the Nineteenth Century', *Technology and Culture*, April 2005.
4. Mel Gorman, 'Electric Lumination in the Franco-Prussian War', *Social Studies of Science*, vol. 7, no. 4 (November 1977).
5. *Electrician*, 24 August 1878, pp.166–7.
6. *New York Sun*, quoted in *The Times*, 8 October 1878.
7. ibid.
8. *The Times*, 24 November 1879.
9. Lady Gwendolen Cecil, *Life of Robert, Marquis of Salisbury*, vol. 3, Hodder & Stoughton, 1931.
10. Report from the Select Committee on Lighting by Electricity 1878–79, Cmd 224.
11. *Liverpool Mercury*, 27 November 1878.

Chapter 2: Swan Shows the Way

1. A memoir of his early life by Swan is included in the book *Sir Joseph Wilson Swan*, by M. E. and K. R. Swan, published by Ernest Benn in 1929.
2. Henry Edmunds, 'Reminiscences of a Pioneer', reprinted from the *M & C Apprentices' Magazine, the Journal of the Apprentices employed at the Works of Messrs Mavor & Coulson, Glasgow*, July, October and December (vol. III, nos. 10, 11, 12, 1919).
3. ibid.
4. ibid.
5. R. S. Taylor, 'Swan's Electric Light at Cragside', in *Papers Presented at the 7th IEE Weekend Meeting on the History of Electrical Engineering* (1979), pp.12–22.
6. Entry for Sir Hiram Maxim in the *Oxford Dictionary of National Biography*.

Chapter 3: La Bataille de Lumières

1. Robert Fox, 'Thomas Edison's Parisian Campaign: incandescent lighting and the hidden face of technology transfer', *Annals of Science*, 53 (1996), 157–93.
2. Quoted from the *Electrician*, 22 October 1881, 'The Electrical Exhibition at Paris', p. 361 and issue of 29 October 1881, p. 377.
3. Swan's Paris letters, quoted in Brian Bowers, 'The Electrical Exhibitions of 1881 and 1882', *Papers Presented at the 8th IEE Weekend Meeting on the History of Electrical Engineering* (1980), and in M. E. and K. R. Swan, *Sir Joseph Wilson Swan*.
4. Edmunds, 'Reminiscences of a Pioneer'.
5. List of awards in the *Electrician*, 29 October 1881.

Chapter 4: The Old Mill on the Stream

1. Patrick Strange, 'The Earliest Electricity Supply: the evidence of Godalming and Chesterfield 1881–84', *Papers Presented at the 6th IEE Weekend Meeting on the History of Electrical Engineering* (1978).

2. Francis Haveron, *The Brilliant Ray*, Godalming Electricity Centenary Celebrations Committee, *c.* 1981.
3. Michael J. Hearn, 'The 125th Anniversary of the Godalming Experimental Electricity Supply System', IET History Technical and Professional Network newsletter, September 2006.
4. *Surrey Advertiser*, 1 October 1881.
5. Quoted in Francis Haveron, *The Brilliant Ray*. The original manuscript of George F. Tanner is in the Local Studies Library at Godalming Museum.
6. *Punch*, 8 October 1881.
7. *Daily Telegraph,* 30 September 1881, quoted in Francis Haveron, *The Brilliant Ray*.

Chapter 5: A Dim View of Electricity

1. Thomas P. Hughes, *Networks of Power*, Johns Hopkins University Press, 1983, pp. 54–7.
2. ibid., pp. 40–45.
3. Brian Bowers, 'Joseph Chamberlain and the First Electric Lighting Act', *Papers Presented at the 11th IEE Weekend Meeting on the History of Electrical Engineering* (1983).
4. Thomas Parke Hughes, 'British Electrical Industry Lag 1882–1888', *Technology and Culture*, vol. 3, no. 1 (Winter 1962), pp. 27–44.
5. Rookes Crompton, *Reminiscences*, Constable, 1928.
6. Thomas Parke Hughes, 'British Electrical Industry Lag 1882–1888', pp. 29–30.
7. Minutes of evidence given by William Siemens to the Select Committee on the Electric Lighting Bill, 19 and 22 May 1882.
8. English Electric Company, *A Collection of Letters to Sir Charles William Siemens, 1823–1883*, 1953.
9. Evidence given by William Siemens to the Select Committee on the Electric Lighting Bill, 19 May 1882.

Chapter 6: A Pre-Raphaelite Power Station

1. Christopher Newall, *The Grosvenor Gallery Exhibitions: change and continuity in the Victorian art world*, Cambridge University Press, 1995.
2. *The Survey of London*, general editor F. H. W. Sheppard, vol. 40, *The Grosvenor Estate in Mayfair*, chapter 3, 'The Aeolian Hall' (formerly Grosvenor Gallery).
3. Rob Cochrane, *Pioneers of Power: the story of the London Electric Supply Corporation*, London Electricity Board, 1987.
4. For the intimate details of Sebastian di Ferranti's early life and career I am indebted to Professor John Wilson's *Ferranti: a history, vol. 1, Building a Family Business, 1882–1975*, Carnegie Publishing, 2000. Professor Wilson's account corrects some of the errors in earlier accounts of Ferranti's background.
5. Rookes Crompton, *Reminiscences*.
6. Robert Hammond, *The Electric Light in Our Homes*, Frederick Warne, 1884.
7. Entry for Sir Francis Arthur Marindin in the *Oxford Dictionary of National Biography*.
8. H. H. Ballin, *The Organisation of Electricity Supply in Great Britain*, Electrical Press, 1946.
9. Leader in *The Times,* 20 May 1889.
10. Rookes Crompton, *Reminiscences*.
11. *Daily News*, 26 September 1889.

Chapter 7: A Tale of Two Cities

1. This chapter is based on J. D. Poulter's *An Early History of Electricity Supply*, published by Peregrinus in 1986.
2. R. H. Parsons, *Early Days of the Power Station Industry*, Cambridge University Press, 1939.
3. 'Municipal Electricity Works', paper read before the Incorporated Association of Municipal and County Engineers, West Bromwich, 1893 (pamphlet in IEE Library).
4. Ian Charles Byatt, *The British Electrical Industry 1875–1914*, Clarendon Press, 1979.

5. Brian Bowers, *Lengthening the Day*, Oxford University Press, 1998, and Maureen Dillon, *Artificial Sunshine*, National Trust, 2002.

Chapter 8: The Americans Ride In

1. *Leeds Mercury*, 19 January 1892.
2. *Australian Dictionary of Biography*, online edition (though American, Train made a fortune in the Australian Gold Rush in the 1850s).
3. Vesey Knox, 'The Economic Effects of the Tramways Act of 1870', *Economic Journal*, vol. 11, no. 44, December 1901, pp. 492–510.
4. Dugald C. Jackson, 'Frank Julian Sprague 1857–1934', *Scientific Monthly*, vol. 57, no. 5, November 1943, pp. 431–41.
5. Carl W. Condit, 'The Pioneer Stage of Railroad Electrification', *Transactions of the American Philosophical Society*, New Series, vol. 67, no. 7 (1977).
6. Vesey Knox, 'The Economic Effects of the Tramways Act of 1870'.
7. Eric Jackson-Stevens, *100 Years of British Electric Tramways*, David & Charles, 1985; T. C. Barker and M. Robbins, *A History of London Transport: passenger travel and the development of the metropolis, vol. 2: The Twentieth Century to 1970*, Allen & Unwin, 1974.
8. T. S. Lascelles, *The City and South London Railway*, Oakwood Press, Lingfield, 1955.
9. R. H. Thurston, 'American Electricians in London', *Science*, New Series, vol. 12, no. 305, 2 November 1900, pp. 689–90.
10. 'Electrical Railways for London – Schemes for the New Session: New Features in an Old Puzzle – A "Boom" in Tubes', *Daily News*, 19 December 1900.
11. Sidney I. Roberts, 'Portrait of a Robber Baron: Charles T. Yerkes', *Business History Review*, vol. 35, no. 3 (Autumn 1961), pp. 344–71.
12. Robert Forrey, 'Charles Tyson Yerkes: Philadelphia-Born Robber Baron', *Pennsylvania Magazine of History and Biography*, vol. 99, no. 2, April 1975, pp. 226–41.

Chapter 9: A Very British Invention

1. Quoted in Rollo Appleyard, *Charles Parsons, His Life and Work*, Constable, 1933.

2. This account of Parsons's life and work is based largely on Rollo Appleyard, *Charles Parsons*.
3. R. H. Parsons, *Early Days of the Power Station Industry*, Cambridge University Press, 1939.
4. Quoted in J. Turnbull, 'The Hon Sir Charles Parsons and Electricity Supply', in *Papers Presented at the 7th IEE Weekend Meeting on the History of Electrical Engineering* (1979), pp. 26–41.

Chapter 10: Electrical Messiahs

1. Letter in the IEE Archive, Merz collection, UK0108 NAEST 091.
2. ibid.
3. Thomas P. Hughes describes the work of Merz and McLellan in detail in *Networks of Power*, Johns Hopkins University, 1983.
4. John Rowland, *Progress in Power: the contribution of Charles Merz and his associates to sixty years of electrical development 1899–1959*, Newman Neame, 1960. See also entry in the *Oxford Dictionary of National Biography*.
5. Autobiography of Charles Merz in the IEE Archives, UK0108 NAEST 091/2 1936.
6. ibid.
7. ibid.

Chapter 11: Electrifying London

1. 'Electricity in London: Notting Hill', *Daily News*, 1 June 1891.
2. 'The Electric Lighting of London No. 6: Clubland', *Daily News*, 8 October 1889.
3. 'The Electric Lighting of London No. 7', *Daily News*, 31 October 1889.
4. 'The Electric Lighting of London No. 5', *Daily News*, 7 October 1889.
5. 'The Electric Lighting of London No. 7'.
6. Mrs J. E. H. Gordon, *Decorative Electricity*, Sampson Low & Company, 1891.
7. W. H. Onken Jr, 'Electrical Development in England', *Electrical World*, 82, 1923.
8. This account of Insull and his relationship with Merz is taken from Thomas

P. Hughes, *Networks of Power*, and Forrest McDonald's biography, *Insull*, University of Chicago Press, 1962.
9. Quoted in John Rowland, *Progress in Power*.

Chapter 12: The Legacy of War

1. Chauncy D. Harris, 'Electricity Generation in London, England', *Geographical Review*, vol. 31, no. 1, January 1941.
2. Arnold B. Gridley and Arnold H. Human, 'Electric Power Supply During the Great War', *Journal of the Institute of Electrical Engineers*, vol. 57, 1919.
3. ibid.
4. *The Times*, 30 May 1918.
5. Report of the Committee appointed to review the National Problem of the Supply of Electrical Energy, 1926.
6. Orren C. Hormell, 'Ownership and Regulation of Electric Utilities in Great Britain', *Annals of the American Academy of Political and Social Science*, vol. 159, part 1, January 1932, pp. 128–39.
7. Leslie Hannah, *Electricity before Nationalisation*, Macmillan, 1979.

Chapter 13: The Quick Perspective of the Future

1. Leslie Hannah, *Electricity before Nationalisation*, and Rob Cochrane, *Power to the People: the story of the National Grid*, Newnes, 1985.
2. Bill Luckin, *Questions of Power: electricity and environment in inter-war Britain*, Manchester University Press, 1990.
3. Leslie Hannah, *Electricity before Nationalisation*.
4. 'Pylons on the Downs', *The Times*, 1 November 1929.
5. 'Pylons on the Downs', *The Times*, 6 November 1929.
6. B. Donoughue and G. W. Jones, *Herbert Morrison: portrait of a politician*, Weidenfeld & Nicolson, 1973.
7. Stephen Spender, *Poems*, Faber and Faber, 1933.
8. *The Times*, 5 September 1933.
9. ibid.
10. Bill Luckin, *Questions of Power*.

11. *The Times*, 6 September 1933.
12. ibid.
13. Leslie Hannah, *Electricity before Nationalisation*.
14. On assisted wiring schemes, see D. J. Bolton, 'Report on the Supply of Electricity in Great Britain', *Economic Journal*, vol. 47, no. 185, March 1937, and Melvin G. de Chazeau, 'The Rationalisation of Electricity Supply in Great Britain', *Journal of Land and Public Utility Economics*, vol. 10, no. 4, November 1934.
15. Eileen Murphy, *The Lure of the Grid*.

Chapter 14: Wired Women and the All-electric Home

1. Peggy Scott, *An Electrical Adventure*, Electrical Association for Women, 1934.
2. Carroll Pursell, 'Domesticating Modernity: the Electrical Association for Women 1924–86', *British Journal of the History of Science*, vol. 32, no. 1, March 1999, pp. 47–67.
3. Entry for Caroline Haslett in the *Oxford Dictionary of National Biography*, and Bill Luckin, *Questions of Power*, chapter 3.
4. G. Z. de Ferranti, *The Life and Letters of Sebastian Ziani di Ferranti* (written with Richard Ince), Williams & Norgate, 1934.
5. 'The Electrical Equipment of Baslow Hall', *Electrical Review*, 17 June 1927.
6. Peggy Scott, *An Electrical Adventure*.
7. E. E. Edwards, *Report on Electricity in Working Class Homes*, Electrical Association for Women, 1935, and Elizabeth Sprenger and Pauline Webb, 'Persuading the Housewife to use Electricity? An interpretation of the material in the Electricity Council Archives', *British Journal of the History of Science*, vol. 26, no. 1, March 1933.
8. Adrian Forty, *Objects of Desire: design and society 1750–1980*, Thames & Hudson, 1986.
9. Christina Hardyment, *From the Mangle to the Microwave*, Polity Press, 1988.
10. Gavin Weightman and Steve Humphries, *The Making of Modern London*, Sidgwick & Jackson, 1984.
11. See Leslie Hannah, *Electricity before Nationalisation*, p. 210, n. 92, for international comparisons of electricity connections.
12. Rob Cochrane, *Power to the People*.

Chapter 15: Battersea and the Barrage Balloon

1. 'Power Stations in Cities', *The Times*, 9 April 1929.
2. Battersea Power Station Memorandum by the Minister of Transport, National Archives CAB/24/203.
3. Bill Luckin, *Questions of Power*, chapter 8.
4. War Period Report of the Electricity Commissioners covering the six years 1st April 1939 to 31 March 1946, HMSO.
5. J. Hacking and J. D. Peattie, 'The British Grid System in Wartime', *Journal of the Institute of Electrical Engineers*, vol. 94, 1947.
6. ibid.
7. IEE archives, Papers of Dame Caroline Haslett, UK010 NAEST Collection 033.

Chapter 16: The Snow Blitz

1. Leslie Hannah, *Electricity before Nationalisation.*
2. Alex J. Robertson, *The Bleak Midwinter: 1947*, Manchester University Press, 1987.
3. Leslie Hannah, *Engineers, Managers and Politicians: the first fifteen years of nationalised electricity supply in Britain*, Macmillan, 1982.
4. Walter Citrine, *Two Careers*, Hutchinson, 1967.
5. Letter from Philip Noel-Baker, 'Fewer Power Cuts', *The Times*, 1 March 1952.
6. Letter from Gerald Nabarro, 'A National Fuel Policy', *The Times*, 5 March 1952.
7. Rob Cochrane, *Power to the People.*
8. Peter Brimblecombe, *The Big Smoke: a history of air pollution in London since medieval times*, Methuen, 1987.
9. Committee on Air Pollution Report, Parliamentary Papers 1954, Cmd 9322.
10. Sue Bowden and Avner Offer, 'Household Appliances and the Use of Time: the United States and Britain since the 1920s', *Economic History Review*, Vol. 47, no. 4, pp. 725–48.

Chapter 17: Power from the Glens

1. This can be found online at http://www.robertburns.org/works/184.shtml.
2. Letter to *The Times*, 1 August 1895.
3. Letter to *The Times*, 'The Falls of Foyers: Mr Ruskin's Opinion', 16 September 1895.
4. Letter to *The Times*, 14 August 1895.
5. Peter L. Payne, *The Hydro*, Aberdeen University Press, 1988.
6. K. J. Lea, 'Hydro-Electric Power Generation in the Highlands of Scotland', *Transactions of the Institute of British Geographers*, no. 46, March 1969, pp. 155–65.
7. Final report of the Board of Trade Departmental Committee on water-power resources (the Snell Report), 1921.
8. Peter L. Payne, *The Hydro*.
9. Report of the Committee on Hydro-electricity Development in Scotland, Cmd 6406, 1942.
10. ibid.
11. Peter L. Payne, *The Hydro*.

Chapter 18: The Promise of Calder Hall

1. *News Chronicle*, 17 October 1956.
2. ibid.
3. Margaret Gowing, *Britain and Atomic Energy 1939–1945*, Macmillan, 1964.
4. Lorna Arnold, *Britain and the H-Bomb*, Palgrave, 2000.
5. *The Times*, leader comment on Monte Bello, 24 October 1952.
6. Roger Williams, *The Nuclear Power Decisions: British policies 1953–78*, Croom Helm, 1980.
7. *The Times*, 18 October 1956.
8. ibid.
9. Richard Green, 'The Cost of Nuclear Power Compared with Alternatives to the Magnox Programme', *Oxford Economic Papers*, New Series, vol. 47, no. 3 (July 1995).

10. Lorna Arnold, *Windscale 1957: anatomy of a nuclear accident*, Macmillan, 1992.
11. 'Milk from Farms near Windscale Stopped', *The Times*, 14 October 1957.
12. Film in the National Archives, 1956, Central Office of Information and British Council.

Chapter 19: A Standard of Living

1. *The Times*, display advertisement, 25 January 1961.
2. *The Times*, display advertisement, 21 February 1961.
3. Sue Bowden and Avner Offer, 'Household Appliances and the Use of Time', and Brian Bowers, 'There's a Lot of It About – I Think', *Papers Presented at the 9th IEE Weekend Meeting on the History of Electrical Engineering* (1981), pp. 28–36.
4. 'Lure of TV Raises Demand for Electricity', *The Times*, 23 May 1961.
5. Ann Oakley, *The Sociology of Housework*, Martin Robinson, 1974.
6. Leslie Hannah, *Engineers, Managers and Politicians.*
7. 'Cross Channel Power Link', Central Electricity Generating Board, 1961, and *Electricity Supply in the United Kingdom: a chronology from the beginnings of the industry to 31 December 1985*, Electricity Council, 1973.
8. 'Channel Power Cable in Use', *The Times*, 5 October 1961.
9. Roger Williams, *The Nuclear Power Decisions*, chapter 8, 'The Quarrel with Coal'.
10. Dieter Helm, *Energy, the State and the Market: British energy policy since 1979*, Oxford University Press, 2003.

Chapter 20: Back to the Future

1. 'Stockton Attacks Thatcher Policies', *The Times*, 9 November 1985.
2. Mike Parker, 'Effects on Demands for Solid Fuels', in J. Surrey (ed.), *The British Electricity Experiment: privatisation, the record, the issues, the lessons*, Earthscan, 1996.
3. Dieter Helm, *Energy, the State and the Market.*
4. Steve Thomas, 'The Privatisation of the Electricity Supply Industry', in J. Surrey (ed.), *The British Electricity Experiment.*
5. Price, Terence, *Political Electricity: what future for nuclear energy?*, Oxford University Press, 1990.

6. Sir Frank Layfield, Sizewell B public inquiry report by HMSO, 1987.
7. Mark Winskel, 'When Systems are Overthrown: the "dash for gas" in the British electricity supply industry', *Social Studies of Science*, vol. 32, no. 4, August 2002, pp. 563–98. Also Mike Parker, *Effects on Demands for Solid Fuels*.
8. Cecil Parkinson, *Right at the Centre*, Weidenfeld & Nicolson, 1992.
9. Cabinet Office, 'The Road to 2010. Addressing the nuclear question in the twenty first century', Cmd 7675.

INDEX